Glencoe McGraw-Hill

Study Notebook

Pre-Algebra

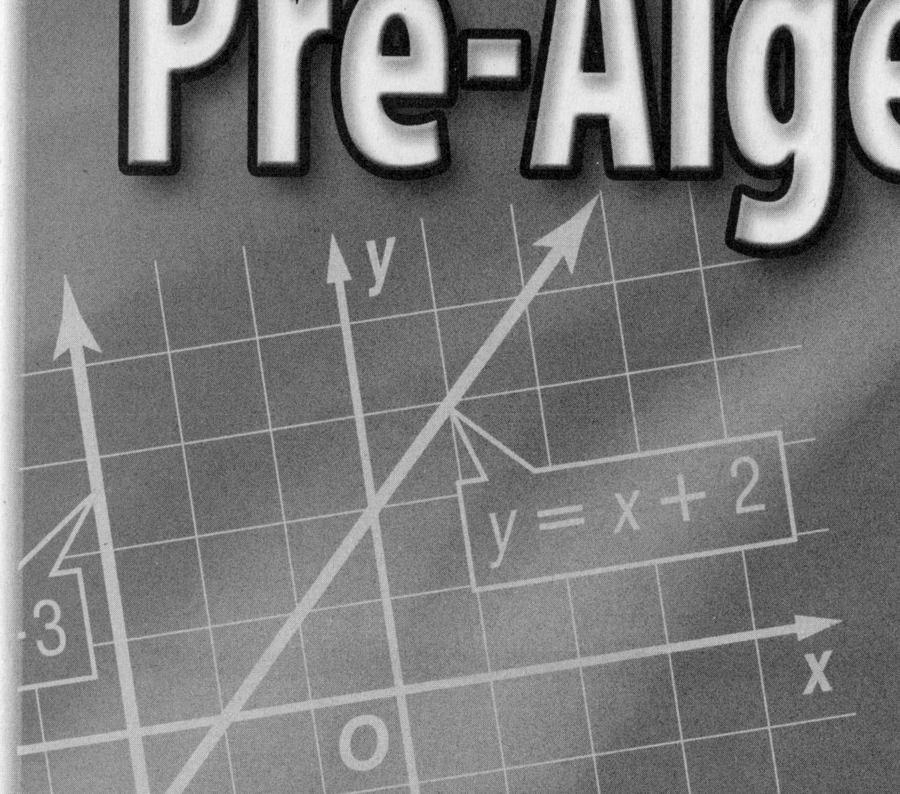

Mc Graw Hill Glencoe

The McGraw·Hill Companies

Glencoe

Send all inquiries to:
Glencoe/McGraw-Hill
8787 Orion Place
Columbus, OH 43240-4027

ISBN: 978-0-07-890748-7
MHID: 0-07-890748-9

Printed in the United States of America

4 5 6 7 8 9 10 MAL 14 13 12 11 10

Contents

Note-Taking Tips

Your notes are a reminder of what you learned in class. Taking good notes can help you succeed in mathematics. The following tips will help you take better classroom notes.

- Before class, ask what your teacher will be discussing in class. Review mentally what you already know about the concept.
- Be an active listener. Focus on what your teacher is saying. Listen for important concepts. Pay attention to words, examples, and/or diagrams your teacher emphasizes.
- Write your notes as clear and concise as possible. The following symbols and abbreviations may be helpful in your note-taking.

Word or Phrase	Symbol or Abbreviation	Word or Phrase	Symbol or Abbreviation
for example	e.g.	not equal	$\neq$
such as	i.e.	approximately	$\approx$
with	w/	therefore	$\therefore$
without	w/o	versus	vs
and	+	angle	$\angle$

- Use a symbol such as a star (★) or an asterisk (*) to emphasis important concepts. Place a question mark (?) next to anything that you do not understand.
- Ask questions and participate in class discussion.
- Draw and label pictures or diagrams to help clarify a concept.
- When working out an example, write what you are doing to solve the problem next to each step. Be sure to use your own words.
- Review your notes as soon as possible after class. During this time, organize and summarize new concepts and clarify misunderstandings.

Note-Taking Don'ts

- Don't write every word. Concentrate on the main ideas and concepts.
- Don't use someone else's notes as they may not make sense.
- Don't doodle. It distracts you from listening actively.
- Don't lose focus or you will become lost in your note-taking.

NAME ______________________ DATE ____________ PERIOD ____________

CHAPTER 1 The Tools of Algebra

Before You Read

Before you read the chapter, respond to these statements.

1. Write an **A** if you agree with the statement.
2. Write a **D** if you disagree with the statement.

Before You Read	Tools of Algebra
	• A variable represents an unknown number or quantity.
	• If the order of numbers multiplied is changed, the product will also change.
	• A coordinate plane has an x- and a y-axis.
	• A scatter plot sometimes shows a trend in the data, but not always.
	• You need an ordered pair with two numbers to plot a point on a coordinate plane.

FOLDABLES® Study Organizer Construct the Foldable as directed at the beginning of this chapter.

Note Taking Tips

- **When you take notes, be sure to describe steps in detail.** Include examples of questions you might ask yourself during problem solving.
- **When searching for the main idea of a lesson, ask yourself, "What is this paragraph or lesson telling me?"** Then make certain you answer the question.

NAME ______________________________ DATE ____________ PERIOD ____________

CHAPTER 1

The Tools of Algebra

Key Points

Scan the pages in the chapter and write at least one specific fact concerning each lesson. For example, in the lesson on variables and expressions, one fact might be that a variable is a letter or symbol used to represent an unknown value. After completing the chapter, you can use this table to review for your chapter test.

Lesson	Fact
1-1 Words and Expressions	
1-2 Variables and Expressions	
1-3 Properties	
1-4 Ordered Pairs and Relations	
1-5 Words, Equations, Tables, and Graphs	
1-6 Scatter Plots	

NAME ______________________________ DATE ____________ PERIOD ____________

1-1 Words and Expressions

What You'll Learn

Skim the text under the *Now* heading. List two things you will learn about in the lesson.

1. ____

____

2. ____

____

Active Vocabulary

New Vocabulary Write the correct term next to each definition.

____________________ ▶ rules to follow when evaluating an expression with more than one operation

____________________ ▶ contains a combination of numbers and operations such as addition, subtraction, multiplication, and division

____________________ ▶ to find the numerical value of an expression

Vocabulary Link *Operation* is a word that is used in everyday English. Find the definition of *operation* using a dictionary. Explain how the English definition can help you remember how *operation* is used in mathematics.

____

____

____

____

____

Main Idea	Details

Translate Verbal Phrases into Expressions

p. 5

Complete the operation of the numerical expressions for each verbal phrase.

1. the number of weeks in 42 days → 42 ☐ 7
2. the difference of 18 and 13 → 18 ☐ 13
3. the quotient of 81 and 9 → 81 ☐ 9
4. the total number of students if there are 7 boys and 11 girls → 7 ☐ 11
5. the total number of tires on 14 cars → 14 ☐ 4
6. the sum of 51 and 39 → 51 ☐ 39
7. the product of 9 and 6 → 9 ☐ 6
8. the cost of 4 candies at \$0.35 each → 4 ☐ 0.35

Order of Operations

p. 6

Complete each step to evaluate 2[(7 + 9) × 3] – 15.

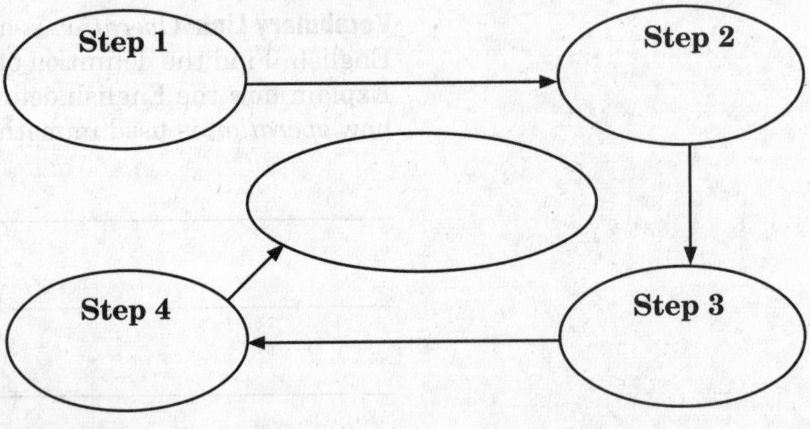

Helping You Remember

One classmate evaluates the expression 4 + 6 ÷ 2 and gets an answer of 5. Another classmate evaluates the same expression and gets an answer of 7. Use the order of operations to explain which answer is correct.

____

____

1-2 Variables and Expressions

What You'll Learn

Skim the lesson. Write two things you already know about variables and expressions.

1. ______________________________

____

2. ______________________________

____

Active Vocabulary

Review Vocabulary Write a numerical expression for each verbal phrase. *(Lesson 1-1)*

18 books shared equally among 6 students ________

a package of 15 pencils minus 3 pencils ________

4 eggs plus 3 eggs ________

New Vocabulary Match the term with its definition by drawing a line to connect the two.

variable	an expression with at least one variable and one operation
defining a variable	branch of mathematics that uses symbols
algebraic expression	a letter or symbol that represents an unknown value
algebra	choosing a variable and the quantity it represents

Main Idea	Details

Algebraic Expressions and Verbal Phrases
pp. 11–12

Describe the steps involved in writing algebraic expressions.

WORDS

↓

VARIABLE

↓

EXPRESSION

Evaluate Expressions
pp. 12–13

Evaluate each expression if $a = 3$, $b = 7$, and $c = 5$.

1. $6c \div 15 = \square$
2. $32 + 4a = \square$
3. $27a - (16 - 3c) = \square$
4. $\dfrac{bc}{a + 2} = \square$
5. $2b - 4a = \square$

Helping You Remember

Variable is a word used in everyday English as well as in mathematics. Write the definition of *variable*. Explain how the English definition can help you remember how *variable* is used in mathematics.

____

____

____

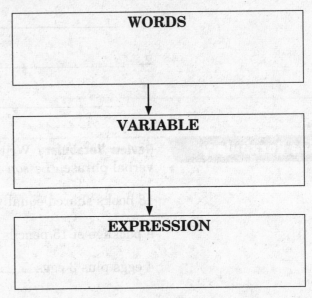

NAME ______________________ DATE __________ PERIOD __________

1-3 Properties

What You'll Learn

Scan Lesson 1-3. List two headings that you would use to make an outline of this lesson.

1. ______________________

2. ______________________

Active Vocabulary

New Vocabulary Write the definition next to each term.

properties ▶ ______________________

counterexample ▶ ______________________

simplify ▶ ______________________

deductive reasoning ▶ ______________________

Vocabulary Link *Simplify* has a non-mathematical meaning as well. Use the word *simplify* in a non-mathematical sentence.

____

Lesson 1-3

Main Idea | **Details**

Properties of Addition and Multiplication

pp. 18–19

Complete the table by writing the definition and an example of each property.

Property	Definition	Example
Commutative Property of Addition		
Associative Property of Multiplication		
Additive Identity		
Multiplication Identity		

Simplify Algebraic Expressions

p. 20

Simplify each expression by filling in the blanks with a variable or number.

1. $(8 + x) + 2$

$\square + \square$

2. $k \times (4 \times 4)$

$\square \times \square$

3. $3 \times (7 \times p)$

$\square \times \square$

4. $9 + (b + 5)$

$\square + \square$

Helping You Remember

In your own words, define counterexample. Tell how it is used in mathematics and why it is important.

____

____

NAME ______________________ DATE ____________ PERIOD ____________

1-4 Ordered Pairs and Relations

What You'll Learn

Skim Lesson 1-4. Predict two things that you expect to learn based on the headings and the Key Concept box.

1. ______________________________

____

2. ______________________________

____

Active Vocabulary

Review Vocabulary Fill in each blank with one correct term. *(Lesson 1-2)*

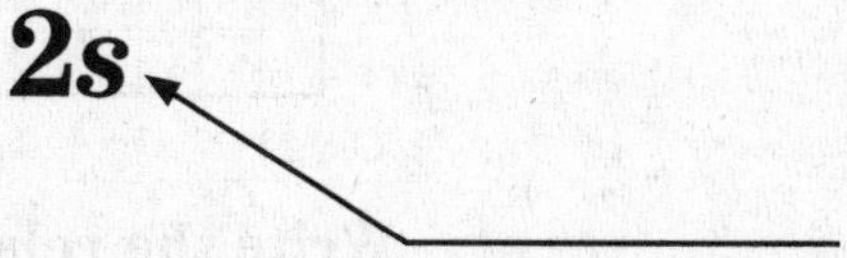

algebraic expression ▶ a(n) ____________ with at least one ____________ and one ____________

New Vocabulary Label the diagram with the correct terms.

orgin ▶

y-axis ▶

x-axis ▶

x-coordinate ▶

y-coordinate ▶

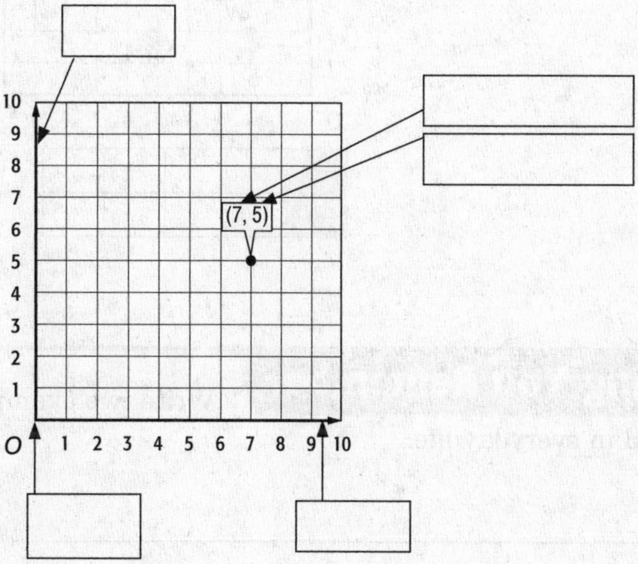

Lesson 1-4

NAME ______________________________ DATE ____________ PERIOD ________

Main Idea	Details

Ordered Pairs
pp. 25–26

Graph each ordered pair on the coordinate plane below.

***A*(6, 4)** ***B*(0, 4)** ***C*(2, 1)** ***D*(5, 0)**

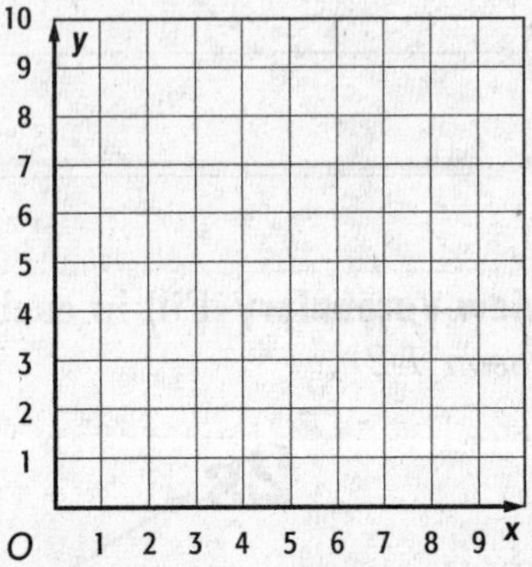

Relations
p. 27

Write the relation as a table. Then write the domain and range.

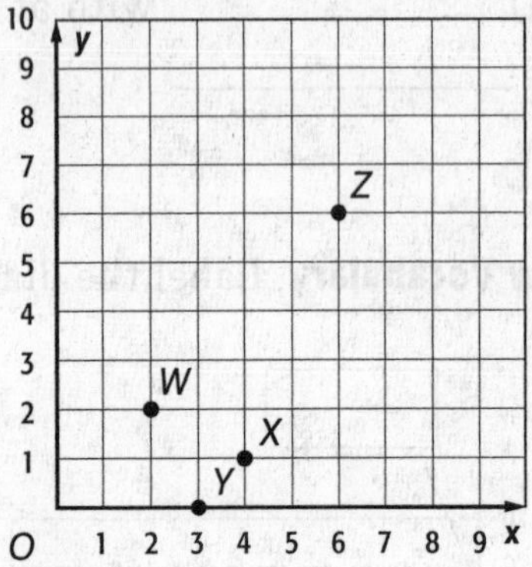

x				
y				

domain: ____________

range: ____________

Helping You Remember

Write two examples of coordinate systems that are used in everyday life.

____

____

1-5 Words, Equations, Tables, and Graphs

What You'll Learn

Skim the Examples for Lesson 1-5. Predict two things you think you will learn about words, equations, tables, and graphs.

1. ______________________________

____

2. ______________________________

____

____

Active Vocabulary

Review Vocabulary Explain how the Additive Identity Property and the Multiplicative Identity Property are the same. *(Lesson 1-3)*

____

____

____

New Vocabulary Fill in each blank with the correct term or phrase.

function ▶ a ____________ where each member of the domain is paired with exactly one member in the ____________

equation ▶ a mathematical ____________ stating that two quantities are ____________

function rule ▶ the ____________ performed on the input in a function to get the ____________

function table ▶ a table that lists the ____________, rule, and the ____________

Main Idea	Details

Represent Functions
p. 33

Complete each function table. Then write the rule for each function.

1. Nancy bought half as many pants as shirts.

Number of shirts	**Input (x)**				
Number of pants	**Output(y)**				

Rule:

2. The recipe calls for 3 times more cups of flour than water.

Cups of water	**Input (x)**				
Cups of flour	**Output (y)**				

Rule:

Multiple Representations
p. 34

Represent the function in three different ways.

For each 1,000 meters in altitude, the temperature, which is 35°C, decreases 6.5°C.

Helping You Remember

Name the four ways that functions can be represented.

____

____

____

1-6 Scatter Plots

What You'll Learn

Scan the text in Lesson 1-6. Write two facts you learned about scatter plots as you scanned the text.

1. ______________________

2. ______________________

Active Vocabulary

Review Vocabulary Complete the table below naming the operation, (addition, subtraction, multiplication, or division) that each verbal phrase represents. *(Lesson 1-1)*

Verbal phrase	Operation
less	
more than	
quotient	
total	
shared equally	
difference	
times	
sum	
product	

New Vocabulary Define the following term from this lesson.

scatter plot ▶ ______________________

Vocabulary Link A *scatter plot* can be used to determine trends between two sets of data. Find the definition of *trend* using a dictionary. Describe how *trend* relates to *scatter plots* using your own words.

____

Main Idea	Details

Construct Scatter Plots
p. 40

Compare and contrast the characteristics of scatter plot and a graphical representation of a function.

Scatter plot	Graph of Function

Analyze Scatter Plots
pp. 41–42

Draw a scatter plot that shows each relationship.

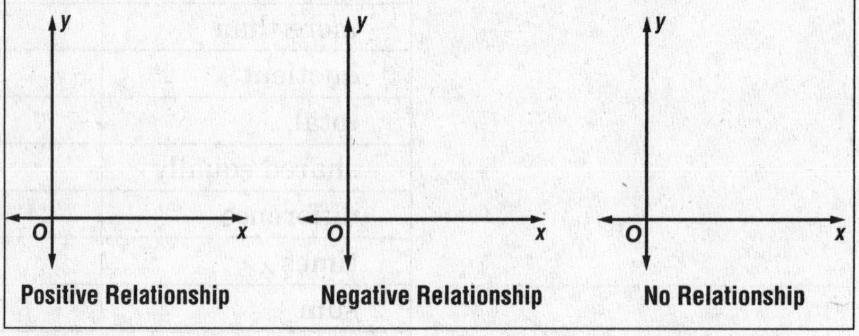

Helping You Remember

Describe three real life situations between quantities where the relationship is a positive relationship, a negative relationship, and no relationship.

____

____

____

____

____

NAME ______________________ DATE ____________ PERIOD ________

CHAPTER 1 The Tools of Algebra

Tie It Together

Complete the table with an example from the chapter.

Property	Symbols	Example(s)
Commutative Property		
Associative Property		
Additive Identity		
Multiplicative Identity		
Multiplicative Property of Zero		

Complete the graphic organizer with a term from the chapter.

{0, 1, 2, 3}

{2, 5, 8, 11}

{(0, 2), (1, 5), (2, 8), (3, 11)}

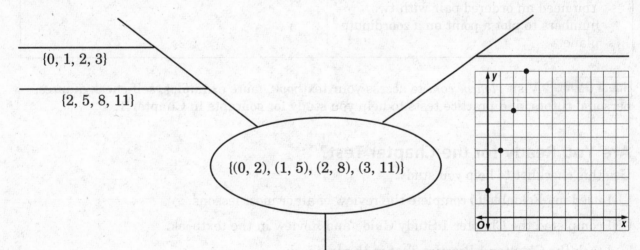

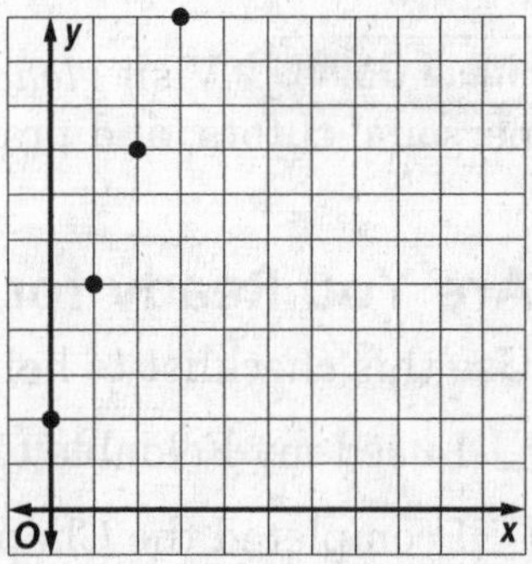

triple a number and add two

Input (x)	0	1	2	3
Output (y)	2	5	8	11

$y = 3x + 2$

CHAPTER 1 Tools of Algebra

Before the Test

Now that you have read and worked through the chapter, think about what you have learned and complete the table below. Compare your previous answers with these.

1. Write an **A** if you agree with the statement.
2. Write a **D** if you disagree with the statement.

Tools of Algebra	After You Read
• A variable represents an unknown number or quantity.	
• If the order of numbers multiplied is changed, the product will also change.	
• A coordinate plane has an x- and a y-axis.	
• A scatter plot sometimes shows a trend in the data, but not always.	
• You need an ordered pair with two numbers to plot a point on a coordinate plane.	

Math Online Visit *glencoe.com* to access your textbook, more examples, self-check quizzes, personal tutors, and practice tests to help you study for concepts in Chapter 1.

Are You Ready for the Chapter Test?

Use this checklist to help you study.

- ☐ I used my Foldable to complete the review of all or most lessons.
- ☐ I completed the Chapter 1 Study Guide and Review in the textbook.
- ☐ I took the Chapter 1 Practice Test in the textbook.
- ☐ I used the online resources for additional review options.
- ☐ I reviewed my homework assignments and made corrections to incorrect problems.
- ☐ I reviewed all vocabulary from the chapter and their definitions.

Study Tips

- Make a calendar that includes all of your daily classes. Besides writing down all assignments and due dates, include in your daily schedule time to study, work on projects, and review notes you took during class that day.

NAME _______________ DATE _______ PERIOD _______

CHAPTER 2

Operations with Integers

Before You Read

Before you read the chapter, respond to these statements.

1. Write an **A** if you agree with the statement.

2. Write a **D** if you disagree with the statement.

Before You Read	Operations with Integers
	• A negative number is less than 0.
	• Every number has one absolute value.
	• Negative numbers can not be used in division problems.
	• When a number is added to its opposite, the sum is zero.
	• The difference of two negative numbers is a negative number.

FOLDABLES® Study Organizer Construct the Foldable as directed at the beginning of this chapter.

Note Taking Tips

- **When you take notes, include definitions of new terms, explanations of new concepts, and examples of problems.**
- **At the end of each lesson, write a summary of the lesson, or write in your own words what the lesson was about.**

NAME ______________________ DATE __________ PERIOD __________

CHAPTER 2

Operations with Integers

Key Points

Scan the pages in the chapter and write at least one specific fact concerning each lesson. For example, in the lesson on integers and absolute value, one fact might be that a positive number is a number greater than zero. After completing the chapter, you can use this table to review for your chapter test.

Lesson	Fact
2-1 Integers and Absolute Value	
2-2 Adding Integers	
2-3 Subtracting Integers	
2-4 Multiplying Integers	
2-5 Dividing Integers	
2-6 Graphing in Four Quadrants	
2-7 Translations and Reflections on the Coordinate Plane	

NAME ________________________ DATE ____________ PERIOD ____________

2-1 Integers and Absolute Value

What You'll Learn

Skim the Examples for Lesson 2-1. Predict two things you think you will learn about integers and absolute value.

1. ______________________________________

2. ______________________________________

____

Active Vocabulary

New Vocabulary Match the term with the correct definition by drawing a line and connecting the two.

negative number	a comparison of numbers or quantities using < or >
positive number	the distance a number is from zero on a number line
integers	a number less than zero
coordinate	the counting numbers, their opposites, and zero
inequality	a number that corresponds to a point on a number line or graph
absolute value	a number greater than zero

Vocabulary Link List three examples of how *negative numbers* are used in everyday life.

____

____

____

Lesson 2-1

Main Idea	Details
Compare and Order Integers pp. 61–62	**Fill in the blank with <, >, or = to make each numerical sentence true.** **1.** –19 ☐ –17 **2.** 0 ☐ –3 **3.** –1 + –3 ☐ –4 **4.** –7 ☐ –10 – 17 **5.** 1 – 6 ☐ 2 – 4
Absolute Value p. 63	**Graph |–5| on a number line. Write its value on the line below your number line. Then explain how you used a number line to find the absolute value of –5.**

Helping You Remember

Absolute is a word used in the English language. Find a definition of *absolute* in a dictionary. Write the definition that most closely relates to mathematics. Explain how the definition you wrote down can help you remember the meaning of *absolute value* in mathematics.

NAME ______________________ DATE __________ PERIOD __________

2-2 Adding Integers

What You'll Learn

Skim Lesson 2-2. Predict two things you expect to learn based on the headings and the Key Concept box.

1. ______________________________

2. ______________________________

____

Active Vocabulary

Review Vocabulary Label the diagram with the correct terms. *(Lesson 2-1)*

positive numbers ▶

negative numbers ▶

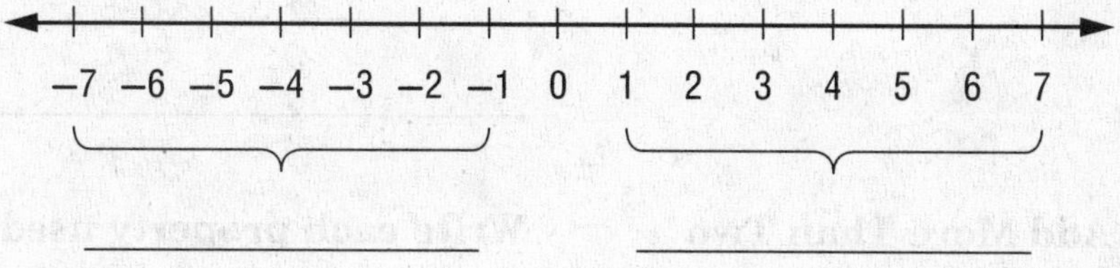

New Vocabulary Fill in each blank with the correct term or phrase.

opposites ▶ two ____________ with the same ______________ but different __________

additive inverse ▶ an __________ and its __________

Vocabulary Link *Opposites* can have non-mathematical meanings as well. Name the opposite of the terms listed.

up ______________________

on ______________________

day ______________________

hot ______________________

boy ______________________

south ______________________

Main Idea	Details
Add Integers pp. 69–71	**Model the addition sentence 3 + (–4) on a number line. Write the sum on the line under your model. Then explain in words how you used the number line to find the sum.**
Add More Than Two Integers p. 72	**Write each property used to simplify the expression.**

$5 + (-7) + 2$

$= 5 + 2 + (-7)$ ____________________

$= (5 + 2) + (-7)$ ____________________

$= 7 + (-7)$ Simplify.

$= 0$ ____________________

Helping You Remember

Suppose that one of your friends was absent from math class the day you learned to add integers. Write an explanation to your friend about how to add integers with the *same signs*. Then explain how to add integers with *different signs*.

____

____

____

____

NAME ______________________ DATE __________ PERIOD __________

2-3 Subtracting Integers

What You'll Learn

Scan the text under the *Now* heading. List two things you will learn about in the lesson.

1. ______________________

2. ______________________

Active Vocabulary

Review Vocabulary Define each term. Include two examples in your definitions. *(Lessons 2-1 and 2-2)*

additive inverse ▶ ______________________

integer ▶ ______________________

opposites ▶ ______________________

Vocabulary Link *Integers* are used in everyday life. For each description, write the integer.

four degrees below zero ________

twelve inches long ________

twenty-five feet below sea level ________

fifty dollars overdrawn ________

Main Idea	Details

Subtract Integers
pp. 76–77

Describe how to subtract integers with the same and different signs and how to add integers with the same and different signs.

	Add Integers	Subtract Integers
same sign		
different signs		

Evaluate Expressions
p. 78

Label the following diagram of a substraction sentence. Then write the subtraction sentence and solve.

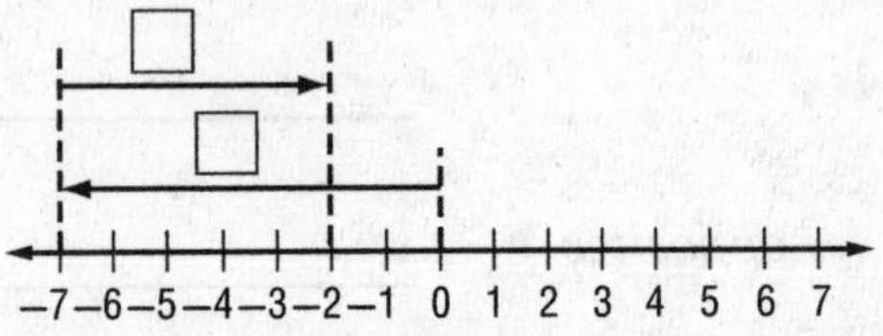

____

Helping You Remember

Write an example for each difference described below. Then use addition to find each difference.

subtract a positive integer from a positive integer ______________________

subtract a positive integer from a negative integer ______________________

subtract a negative integer from a positive integer ______________________

subtract a negative integer from a negative integer ______________________

NAME ______________________ DATE __________ PERIOD __________

2-4 Multiplying Integers

What You'll Learn

Scan the text in Lesson 2-4. Write two facts you learned about multiplying integers.

1. ______________________________

____

____

2. ______________________________

____

Active Vocabulary

Vocabulary Link *Commute* and *associate* are words that are used in everyday English. Find the definition of *commutative* and *associative* using a dictionary. Explain how the English definitions can help you remember how *commutative* and *associative* are used in mathematics.

____

____

____

____

____

____

____

Main Idea	Details
Multiply Integers pp. 83–85	**Fill in the boxes to simplify each expression.**

1. –3(–5) ☐ **2.** –8 × 4 ☐

3. 12 × 10 ☐ **4.** 4(–2) ☐

5. –9(–7) ☐ **6.** –6 · 6 ☐

Algebraic Expressions
p. 85

Simplify the expression given the reason for each step.

–3(12 + *m* + 18)

= ______________ Replace *m* with 50.

= ______________ Commutative Property

= ______________ Simplify inside.

= ______________ Multiply.

Helping You Remember

In your own words, explain why the product of three negative integers is negative. Give an example.

____

____

____

____

____

NAME ______________________ DATE ____________ PERIOD ________

2-5 Dividing Integers

What You'll Learn

Scan the text in Lesson 2-5. Write two facts you learned about dividing integers.

1. ____

____

____

2. ____

____

____

Active Vocabulary

Review Vocabulary Fill in the blank with the correct value. *(Lessons 2-1 and 2-2)*

additive inverse ▶ The *additive inverse* of –8 is ______.

opposite ▶ The *opposite* of 6 is ______.

absolute value ▶ The *absolute value* of |–9| is ______.

New Vocabulary Write the definition next to the term.

mean ▶ ____

____

Vocabulary Link Write two examples of how the mathematical term *mean* is used in everyday life.

____

____

____

Lesson 2-5

NAME ______________________ DATE ____________ PERIOD ____________

Main Idea	Details

Divide Integers
pp. 90–91

Write *positive* or *negative* to identify each quotient.

$-21 \div 7$ $\square$	$-18 \div (-3)$ $\square$
$24 \div 6$ $\square$	$15 \div (-3)$ $\square$

Mean (Average)
p. 92

Evan wanted to make sure his golf score average did not go above 42. He previously scored 44, 38, 33, 47, and 41. Fill in the blanks to solve the equation for *x* that will tell Evan the maximum score he could get and still have an average of 42.

$\frac{44 + 38 + 33 + 47 + 41 + x}{\square} = 42$ There are 6 data items.

$\frac{\square + x}{\square} = 42$ Find the sum of the numerator.

$6\left(\frac{\square + x}{\square}\right) = 42 \times \square$ Eliminate the denominator by multiplying each side by 6.

$\square + x = \square$ Simplify.

$x = \square$ Subtract 203 from each side.

Helping You Remember

Write one example of each quotient described below. Then find the quotient.

dividing a positive integer by a negative integer ____________________

dividing a negative integer by a negative integer ____________________

dividing a negative integer by a positive integer ____________________

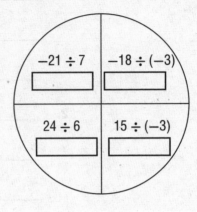

NAME ______________________ DATE __________ PERIOD __________

2-6 Graphing in Four Quadrants

What You'll Learn

Skim the lesson. Write two things you already know about graphing in four quadrants.

1. ______________________________

____

2. ______________________________

____

Active Vocabulary

Review Vocabulary Label the diagram with the correct terms. *(Lesson 1-4)*

origin ▶

y-axis ▶

x-axis ▶

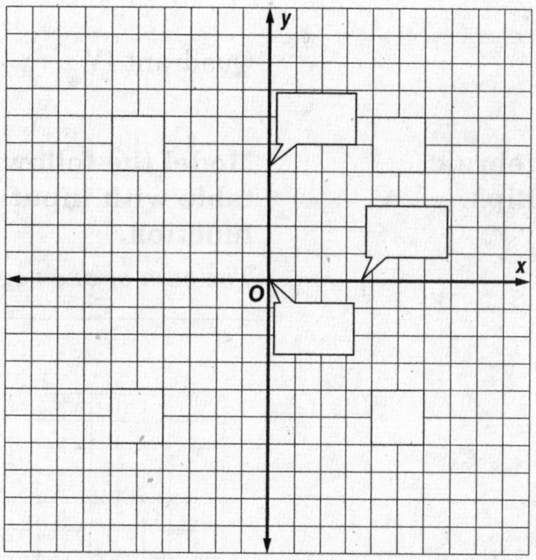

x-coordinate ▶

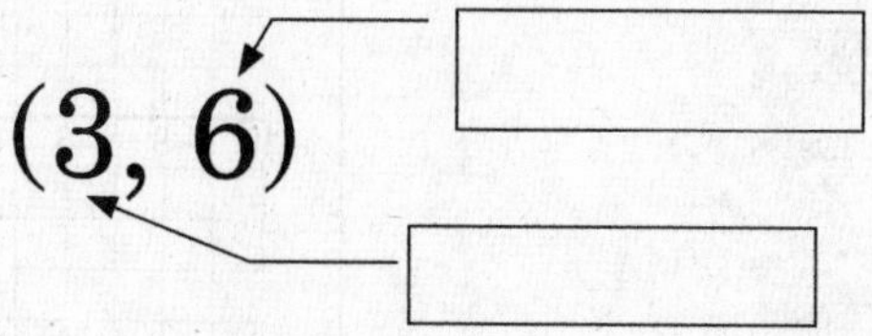

y-coordinate ▶

quadrants ▶ **New Vocabulary** Label the diagram above with the correct quadrant: I, II, III, or IV.

NAME ______________________ DATE __________ PERIOD __________

Main Idea	Details

Graph Points
pp. 96–97

Match the ordered pairs with the correct quadrant by drawing a line to connect the coordinates with the quadrant. Circle coordinates that are not in any quadrant.

Quadrant I	(–3, –3)
	(–2, 0)
Quadrant II	(1, –5)
	(6, 2)
Quadrant III	(–1, 4)
Quadrant IV	(0, 0)

Graph Algebraic Relationships
p. 97

Model the following function by creating a function table with input and output values. Then graph the function.

The sum of one negative and one positive number is 3.

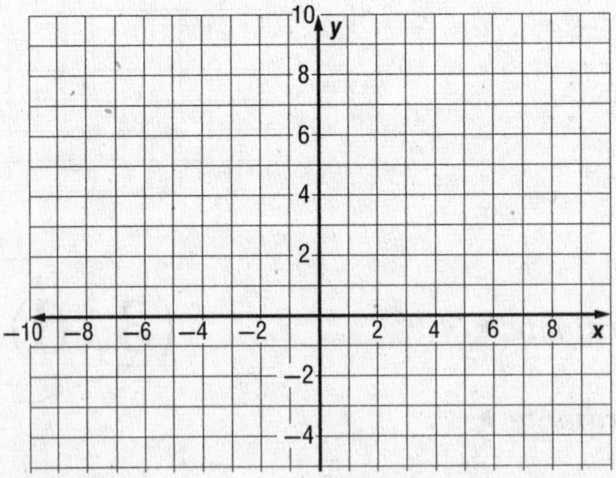

Helping You Remember

Draw a coordinate grid with points to represent your classroom and where your classmates sit. Explain how to name the location of your classmates.

NAME ______________________________ DATE ____________ PERIOD ____________

2-7 Translations and Reflections on the Coordinate Plane

What You'll Learn

Skim Lesson 2-7. Predict two things that you expect to learn based on the headings and the Key Concept box.

1. ____

____

2. ____

____

Active Vocabulary

New Vocabulary Write the correct term next to each definition.

____________________ ▶ a transformation where an original figure is flipped over a line of symmetry

____________________ ▶ a transformation where an original figure moves the same distance in the same direction without turning

____________________ ▶ a line of reflection

____________________ ▶ an operation that maps an original geometric figure onto a new figure

____________________ ▶ a transformed figure

Vocabulary Link *Transform* is a word that is used in everyday English. Find the definition of *transform* using a dictionary. Explain how the English definition can help you remember how *transformation* is used in mathematics.

____

____

____

Main Idea	Details

Transformations
p. 101

Complete the organizer by defining the terms in your own words.

Transformation	
Translation	**Reflection**

Translations and Reflections
pp. 102–103

Compare and contrast *translation* and *reflection*.

	Translation	Reflection
How they are alike		
How they are different		

Helping You Remember

Identify each type of transformation. Then describe in your own words how you know that you are correct.

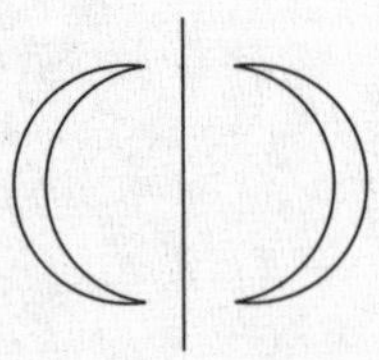

____

____

CHAPTER 2

Operations with Integers

Tie It Together

Complete the graphic organizer with a phrase to help you remember the process.

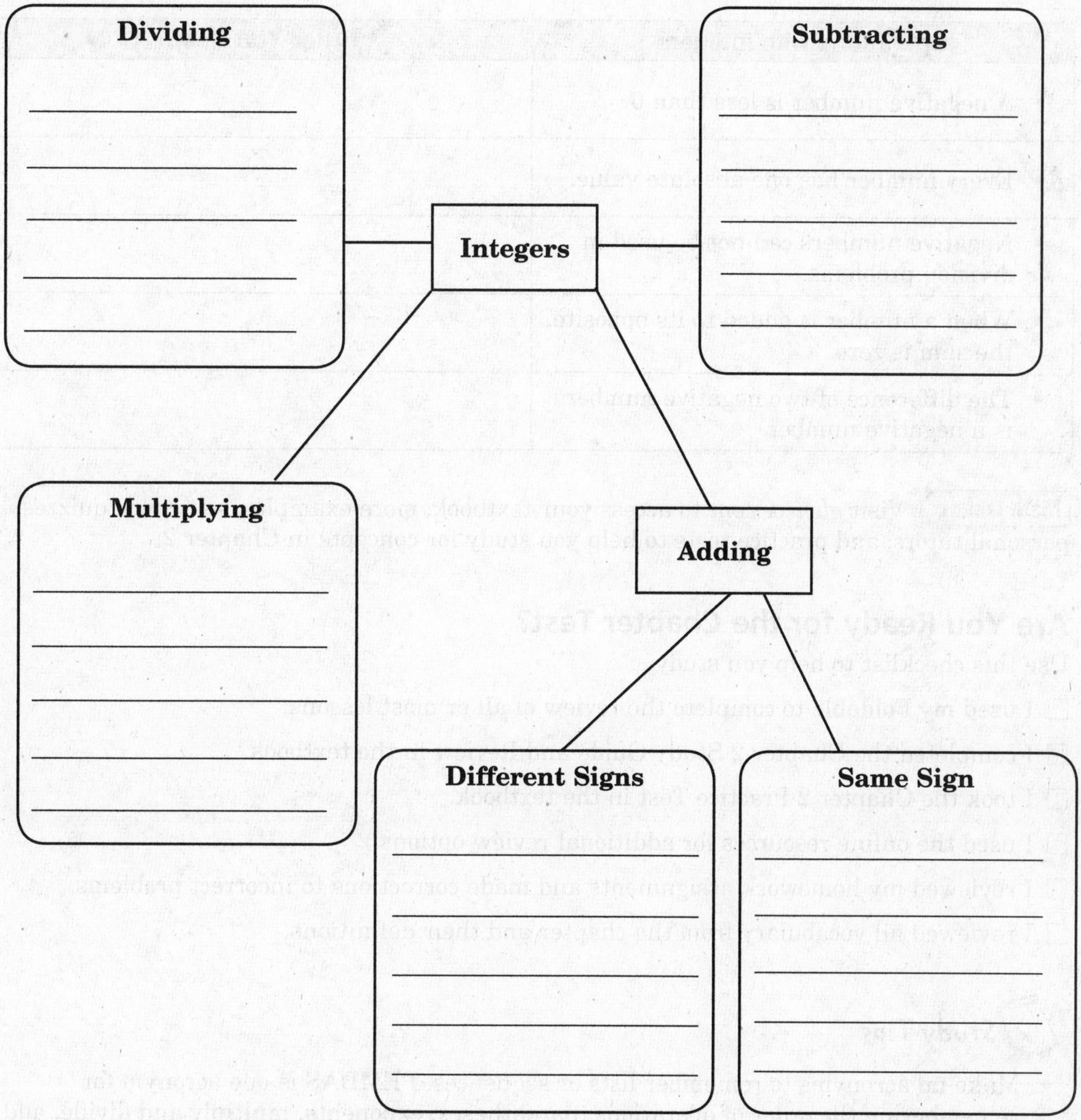

NAME ______________________ DATE __________ PERIOD __________

CHAPTER 2 Operations with Integers

Before the Test

Now that you have read and worked through the chapter, think about what you have learned and complete the table below. Compare your previous answers with these.

1. Write an **A** if you agree with the statement.
2. Write a **D** if you disagree with the statement.

Operations with Integers	After You Read
• A negative number is less than 0.	
• Every number has one absolute value.	
• Negative numbers can not be used in division problems.	
• When a number is added to its opposite, the sum is zero.	
• The difference of two negative numbers is a negative number.	

Math Online Visit *glencoe.com* to access your textbook, more examples, self-check quizzes, personal tutors, and practice tests to help you study for concepts in Chapter 2.

Are You Ready for the Chapter Test?

Use this checklist to help you study.

☐ I used my Foldable to complete the review of all or most lessons.

☐ I completed the Chapter 2 Study Guide and Review in the textbook.

☐ I took the Chapter 2 Practice Test in the textbook.

☐ I used the online resources for additional review options.

☐ I reviewed my homework assignments and made corrections to incorrect problems.

☐ I reviewed all vocabulary from the chapter and their definitions.

Study Tips

- Make up acronyms to remember lists or sequences. PEMDAS is one acronym for remembering the order of operations (parentheses, exponents, multiply and divide, add and subtract). (Please Excuse My Dear Aunt Sally)

NAME ______________________ DATE __________ PERIOD __________

CHAPTER 3

Operations with Rational Numbers

Before You Read

Before you read the chapter, think about what you know about rational numbers. List three things you already know about operations with rational numbers in the first column. Then list three things you would like to learn about them in the second column.

K What I know...	W What I want to find out...

FOLDABLES® Study Organizer Construct the Foldable as directed at the beginning of this chapter.

Note Taking Tips

- **As you read each lesson, list ways the new knowledge has been or will be in your daily life.**
- **When you take notes, record real-life examples of how you can use fractions and decimals such as telling time and making change.**

NAME ______________________ DATE ____________ PERIOD ____________

CHAPTER 3

Operations with Rational Numbers

Key Points

Scan the pages in the chapter and write at least one specific fact concerning each lesson. For example, in the lesson on dividing rational numbers, one fact might be that reciprocals are two numbers whose product is 1. After completing the chapter, you can use this table to review for your chapter test.

Lesson	Fact
3-1 Fractions and Decimals	
3-2 Rational Numbers	
3-3 Multiplying Rational Numbers	
3-4 Dividing Rational Numbers	
3-5 Adding and Subtracting Like Fractions	
3-6 Adding and Subtracting Unlike Fractions	

NAME ______________ DATE __________ PERIOD __________

3-1 Fractions and Decimals

What You'll Learn

Scan the text under the *Now* heading. List two things you will learn about in this lesson.

1. ______________________________

____

2. ______________________________

____

Lesson 3-1

Active Vocabulary

Review Vocabulary Define *inequality* in your own words. *(Lesson 2-1)*

inequality ▶ ______________________________

____

New Vocabulary Match the term with its definition.

bar notation	decimals with a pattern in digits that have no end
terminating decimal	line placed over repeating digits in a decimal
repeating decimal	decimals that divide evenly with no remainder

Vocabulary Link *Terminate* is a word that is used in everyday English. Find the definition of *terminate* using a dictionary. Explain how the English definition can help you remember how *terminate* is used in mathematics.

____

____

____

Main Idea	**Details**

Write Fractions as Decimals

pp. 121–123

Complete the diagram by comparing and contrasting repeating decimals and terminating decimals.

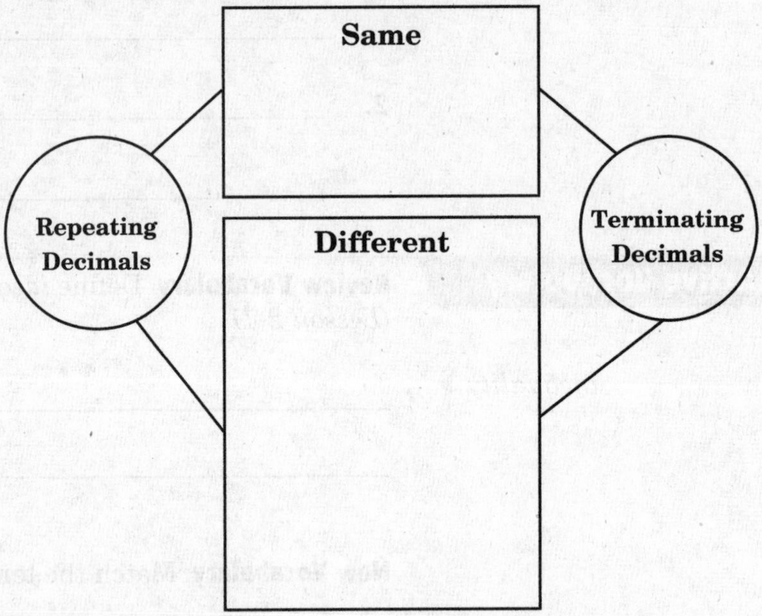

Compare Fractions and Decimals

pp. 123–124

Fill in the blank with <, >, or = to make each numerical sentence true.

1. $\frac{5}{6} \square \frac{2}{3}$ **2.** $-0.36 \square -\frac{1}{3}$

3. $\frac{23}{100} \square \frac{1}{5}$ **4.** $\frac{7}{19} \square \frac{4}{15}$

5. $-\frac{7}{8} \square -\frac{8}{9}$ **6.** $-\frac{1}{5} \square -0.2$

7. $\frac{3}{8} \square \frac{6}{7}$ **8.** $\frac{4}{11} \square \frac{5}{21}$

Helping You Remember

In your own words, explain the difference between 0.6 and $0.\overline{6}$. Which number is greater?

____

____

____

NAME ______________________ DATE __________ PERIOD __________

3-2 Rational Numbers

What You'll Learn

Scan the text in Lesson 3-2. Write two facts you learned about rational numbers as you scanned the text.

1. ______________________________

____

2. ______________________________

____

____

Active Vocabulary

Review Vocabulary Write the definition next to each term. *(Lessons 1-3 and 2-1)*

integers ▶ ______________________________

____

properties ▶ ______________________________

____

New Vocabulary Fill in the blanks with the correct term or phrase.

rational numbers ▶ any __________ that can be written as __________

Vocabulary Link *Rational* is a word used in everyday English. Find the definition of *rational* in a dictionary. Then use the dictionary to find the antonym, or a word that means the opposite, of *rational*.

____

____

____

Lesson 3-2

Main Idea	Details

Rational Numbers
pp. 128–129

Match each repeating decimal with its equivalent fraction.

$0.\overline{3}$	$\frac{14}{33}$
$0.\overline{125}$	$\frac{1}{3}$
$0.\overline{42}$	$\frac{1}{33}$
$0.\overline{03}$	$\frac{7}{9}$
$0.\overline{7}$	$\frac{125}{999}$

Identify and Classify Rational Numbers
p. 130

Complete the diagram by labeling each oval with the correct set of numbers. Use the terms *whole numbers* and *integers*. Then include three examples of whole numbers, integers, and rational numbers.

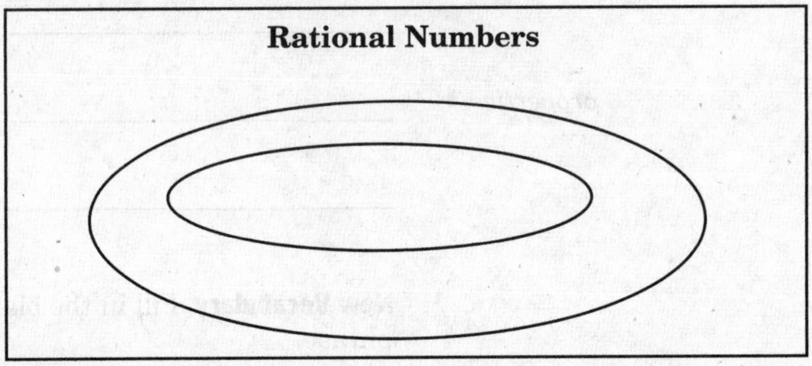

Helping You Remember

Describe the relationship among whole numbers, integers, and rational numbers, in your own words. Give an example of a number that is *not* rational and explain why it is not.

____

____

____

____

NAME ______________________ DATE __________ PERIOD __________

3-3 Multiplying Rational Numbers

What You'll Learn

Skim Lesson 3-3. Predict two things that you learn based on the headings and figures in the lesson.

1. ____

____

2. ____

____

____

Active Vocabulary

Review Vocabulary Write the correct term next to each definition. *(Lessons 2-2 and 3-2)*

______________ ▶ a number less than zero

______________ ▶ the counting numbers, their opposites, and zero

______________ ▶ any number than can be written as a fraction

______________ ▶ a number greater than zero

Vocabulary Link *Multiplication* is the same as repeated addition. In Lesson 2-2 you used a number line to add integers. Explain how you can use 'repeated addition' to demonstrate $4 \cdot \frac{1}{2}$ on a number line.

____

____

____

Lesson 3-3

Main Idea	Details

Multiply Fractions
pp. 134–135

Use the model to find $\frac{3}{5} \times \frac{4}{7}$. Explain your steps on the lines below.

____

____

____

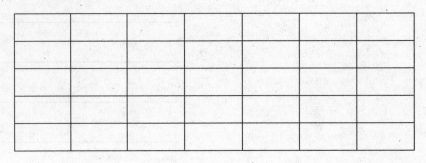

Evaluate Expressions with Fractions
p. 135

Fill in the blanks to find each product in simplest form.

Use $x = \frac{2}{3}$, $y = -\frac{7}{11}$, and $z = \frac{3}{5}$.

1. xy

☐ × ☐ = ☐

2. $-2y$

-2 × ☐ = ☐

3. $\frac{5}{9}z$

$\frac{5}{9}$ × ☐ = ☐

4. xyz

☐ × ☐ × ☐ = ☐

Helping You Remember

Explain in your own words how to find the product of two fractions with a model. What portion of the model is the product?

____

____

____

3-4 Dividing Rational Numbers

What You'll Learn

Scan Lesson 3-4. List two headings you would use to make an outline for this lesson.

1. ______________________________

____

2. ______________________________

____

Active Vocabulary

Review Vocabulary Fill in the blank with the correct term or phrase. *(Lesson 2-2)*.

additive inverse ▶ an ____________ and its opposite

New Vocabulary Write the definition next to each term.

multiplicative inverses ▶ ______________________________

reciprocals ▶ ______________________________

____

____

Vocabulary Link *Reciprocal* is a word that is used in everyday English. Find the definition of *reciprocal* using a dictionary. Explain how the English definition can help you remember how *reciprocal* is used in mathematics.

____

____

____

____

NAME ____________________ DATE __________ PERIOD __________

Main Idea	Details

Divide Fractions
pp. 141–143

Place three division expressions in each section of the Venn diagram.

Quotients Less Than or Equal to 1 | **Quotients Greater Than or Equal to 1**

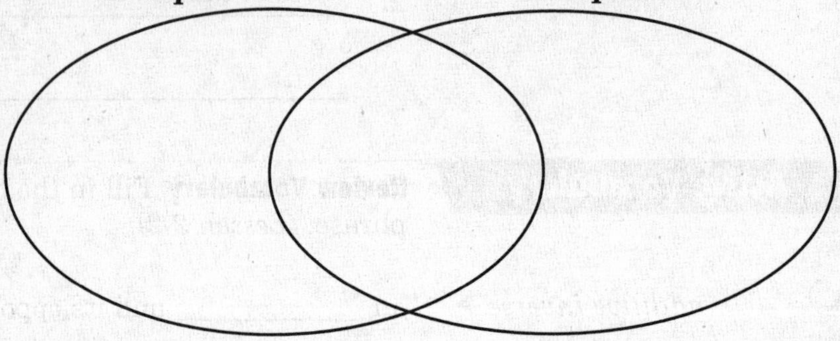

Divide Algebraic Expressions
p. 143

Simplify each expression.

1. $\frac{x^2}{4} \quad \frac{xy}{2}$ ☐

2. $\frac{b}{6ab} \quad \frac{3b}{a}$ ☐

3. $\frac{7}{gh} \quad \frac{5}{4fh}$ ☐

4. $\frac{14x}{xy} \quad \frac{1}{10xy}$ ☐

5. $\frac{q}{12} \quad \frac{n^2}{2}$ ☐

6. $\frac{b}{2d} \quad \frac{2}{9c}$ ☐

Helping You Remember

In your own words, explain how you know whether the quotient of two fractions will be less than 1, equal to 1, or greater than 1.

____

____

____

NAME ______________________ DATE __________ PERIOD __________

3-5 Adding and Subtracting Like Fractions

What You'll Learn

Skim the examples for Lesson 3-5. Predict two things you think you will learn about adding and subtracting like fractions.

1. ______________________________

____

____

2. ______________________________

____

____

Active Vocabulary

Review Vocabulary Write the correct term next to each definition. *(Lesson 2-1).*

______________ ▶ the distance a number is from zero on a number line

______________ ▶ a number less than zero

______________ ▶ the counting numbers, their opposites, and zero

______________ ▶ a number greater than zero

New Vocabulary Write the definition next to the term.

like fractions ▶ ______________________________

____

Vocabulary Link *Like* is a word that is used in everyday English. Find the definition of *like* using a dictionary. Explain how the English definition can help you remember how *like* is used in mathematics.

____

____

Main Idea | **Details**

Add Like Fractions
pp. 147–148

Complete the diagram.

What It Is	What It Is Not
Examples	**Non Examples**

Center: **Like Denominators**

Subtract Like Fractions
pp. 148–150

Fill in the blanks with each difference.

1. $3\frac{3}{5} - 2\frac{2}{5} =$ ☐

2. $8\frac{3}{7} - 5\frac{5}{7} =$ ☐

3. $\frac{1}{9} - 1\frac{8}{9} =$ ☐

4. $\frac{3}{10} - \frac{9}{10} =$ ☐

Helping You Remember

Sketch a model to show each sum or difference.

a. $\frac{3}{10} + \frac{6}{10}$

b. $\frac{6}{7} - \frac{3}{7}$

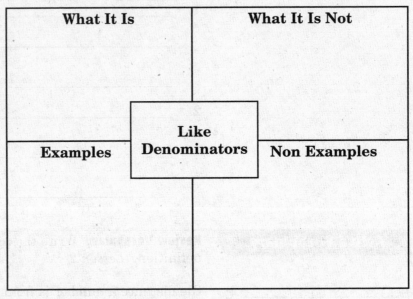

3-6 Adding and Subtracting Unlike Fractions

What You'll Learn

Skim the lesson. Write two things you already know about adding and subtracting unlike fractions.

1. ____

____

2. ____

____

Active Vocabulary

Review Vocabulary Write the definition next to the term. *(Lessons 2-2 and 3-5)*

opposites ▶ ____

____

New Vocabulary Label the diagram with the correct terms.

like fractions ▶

unlike fractions ▶

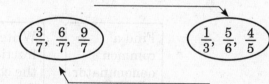

$\frac{3}{7}, \frac{6}{7}, \frac{9}{7}$ ______________

$\frac{1}{3}, \frac{5}{6}, \frac{4}{5}$ ______________

Vocabulary Link *Unlike* can have non-mathematical meanings as well. Give an example of two things that are *unlike* each other. Then give an example of two things that are like each other.

unlike: ____

____

like: ____

____

Main Idea	Details

Add Unlike Fractions
pp. 153–154

Shade each circle to show equivalent fractions for $\frac{1}{3}$ and $\frac{1}{2}$ using the LCD. Then write the addition sentence the model represents.

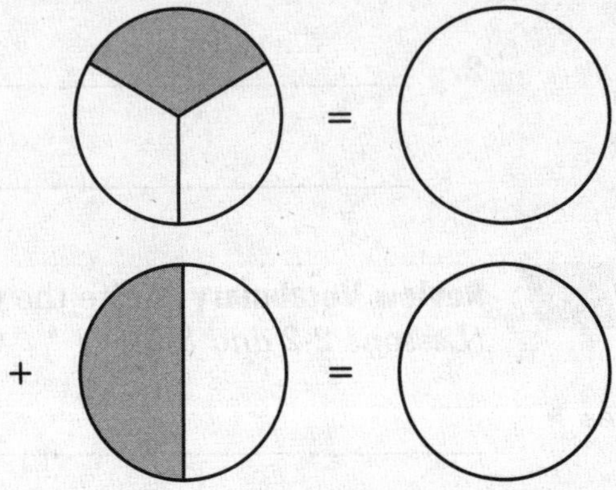

____

Subtract Unlike Fractions
pp. 154–155

Write each step used to find $\frac{5}{6} - \frac{1}{4}$.

Find a common denominator. ☐ → Write each fraction with the common denominator. ☐ → Subtract the numerators, write the difference over the common denominator. ☐

Helping You Remember

Describe two methods that you can use to add $1\frac{1}{3}$ and $3\frac{3}{5}$. Then find the sum.

____

____

CHAPTER 3

Operations with Rational Numbers

Tie It Together

Complete each graphic organizer with a phrase to help you remember the process.

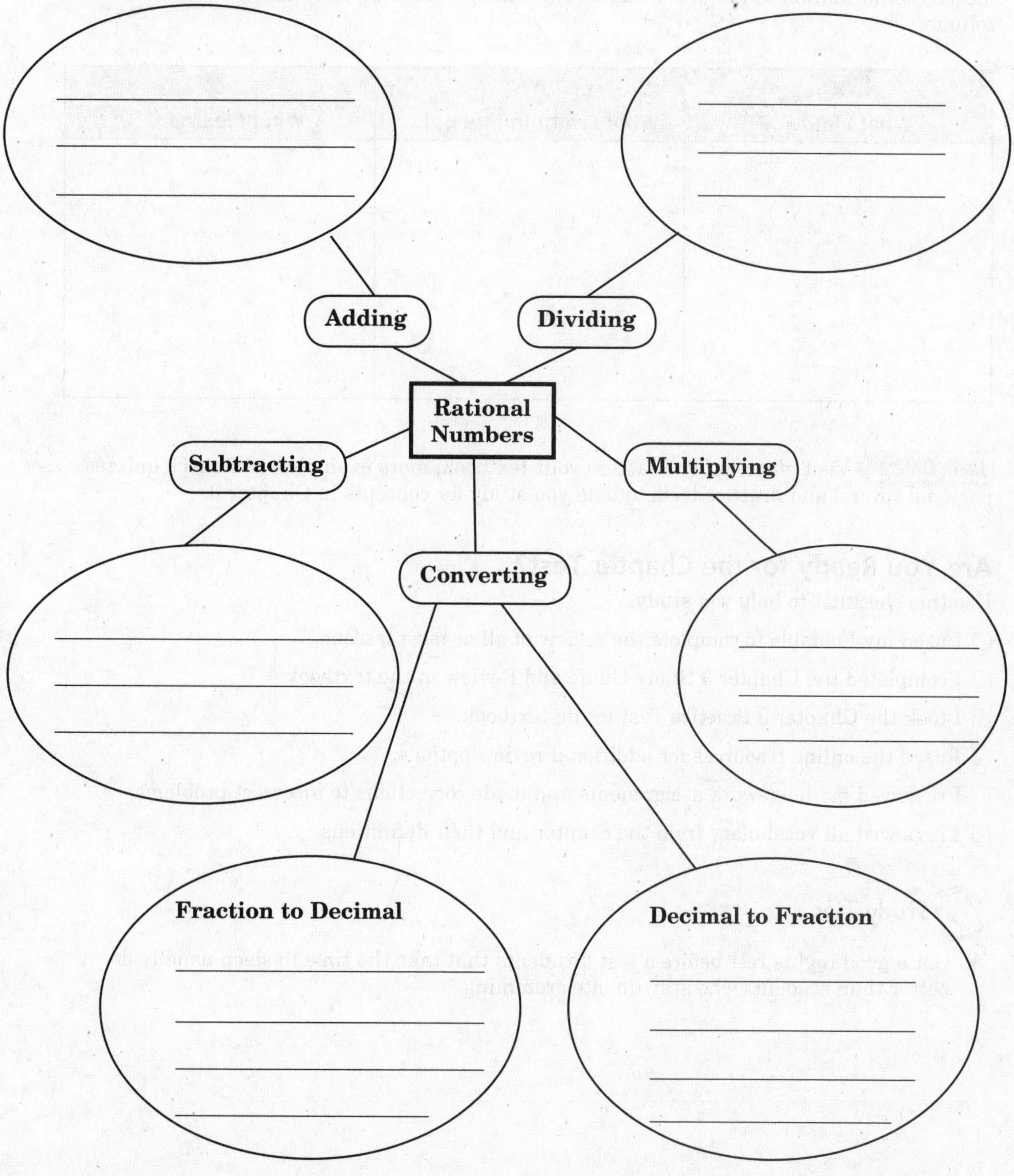

NAME ______________________ DATE ____________ PERIOD ____________

CHAPTER 3 Operations with Rational Numbers

Before the Test

Review the ideas you listed in the table at the beginning of the chapter. Cross out any incorrect information in the first column. Then complete the table by filling in the third column.

K What I know...	W What I want to find out...	L What I learned...

Math Online Visit *glencoe.com* to access your textbook, more examples, self-check quizzes, personal tutors, and practice tests to help you study for concepts in Chapter 3.

Are You Ready for the Chapter Test?

Use this checklist to help you study.

- ☐ I used my Foldable to complete the review of all or most lessons.
- ☐ I completed the Chapter 3 Study Guide and Review in the textbook.
- ☐ I took the Chapter 3 Practice Test in the textbook.
- ☐ I used the online resources for additional review options.
- ☐ I reviewed my homework assignments and made corrections to incorrect problems.
- ☐ I reviewed all vocabulary from the chapter and their definitions.

Study Tips

- Get a good nights rest before a test. Students that take the time to sleep usually do better than students who stay up late cramming.

NAME ______________________ DATE __________ PERIOD __________

CHAPTER 4

Expressions and Equations

Before You Read

Before you read the chapter, think about what you know about expressions and equations. List three things you already know about them in the first column. Then list three things you would like to learn about them in the second column.

K What I know…	W What I want to find out…

FOLDABLES® Study Organizer Construct the Foldable as directed at the beginning of this chapter.

Note Taking Tips

- **When you take notes, listen or read for main ideas.** Then record those ideas for future reference.
- **Write down questions that you have about what you are reading in the lesson.** Then record the answer to each question as you study the lesson.

NAME ______________________ DATE ____________ PERIOD ________

CHAPTER 4

Expressions and Equations

Key Points

Scan the pages in the chapter and write at least one specific fact concerning each lesson. For example, in the lesson on simplifying algebraic expressions, one fact might be that a term without a variable is called a constant. After completing the chapter, you can use this table to review for your chapter test.

Lesson	Fact
4-1 The Distributive Property	
4-2 Simplifying Algebraic Expressions	
4-3 Solving Equations by Adding or Subtracting	
4-4 Solving Equations by Multiplying or Dividing	
4-5 Solving Two-Step Equations	
4-6 Writing Equations	

NAME ______________________ DATE __________ PERIOD __________

4-1 The Distributive Property

What You'll Learn

Skim Lesson 4-1. Predict two things that you expect to learn based on the headings and the Key Concept box.

1. ______________________

____

2. ______________________

____

Active Vocabulary

New Vocabulary Write the definition next to each term.

equivalent expressions ▶ ______________________

____

Distributive Property ▶ ______________________

____

____

Vocabulary Link *Distribute* is a word that is used in everyday English. Find the definition of *distribute* using a dictionary. Explain how the English definition can help you remember how *distributive* is used in mathematics.

____

____

____

____

Lesson 4-1

Main Idea	Details

Numerical Expressions
pp. 171–172

Complete each expression using the Distributive Property.

1. $5(3 + 4) = 5 \cdot 3 + 5 \cdot \square$
2. $6(4 - 1) = 6 \cdot \square - 6 \cdot 1$
3. $2(8 - 7) = 2 \cdot \square - 2 \cdot \square$
4. $3(4 + 9) = \square \cdot 4 + \square \cdot 9$
5. $(2 + 5)8 = 2 \cdot 8 + 5 \cdot \square$
6. $(6 - 3)7 = \square \cdot 7 - \square \cdot 7$

Algebraic Expressions
pp. 172–173

Model the expression $3(x + 2)$. Then model 3 groups of x and 3 groups of 2. Write two equivalent expressions below your model.

Helping You Remember

A classmate states that $3(x + 1) = 3x + 1$. How would you model to your classmate that $3(x + 1) = 3x + 3$ *not* $3x + 1$ Include drawings with your explanation.

____

____

____

NAME ______________________ DATE __________ PERIOD __________

4-2 Simplifying Algebraic Expressions

What You'll Learn

Scan Lesson 4-2. List two headings you would use to make an outline of this lesson.

1. ______________________

____

2. ______________________

____

Active Vocabulary

New Vocabulary Match the term with its definition by drawing a line to connect the two.

coefficient	a term without a variable
constant	each part of an algebraic expression separated by plus and/or minus signs
like terms	terms that contain the same variables
simplest form	the numerical part of a term that contains a variable
simplifying the expression	an algebraic expression that has no like terms and no parentheses
term	creating an equivalent expression with no like terms

Lesson 4-2

Main Idea	Details

Parts of Algebraic Expressions

pp. 178–179

Identify the parts of the algebraic expression below.

$$4x + 9y + 7y - 2x + 5$$

How many terms are there in the expression? ____________

How many sets of like terms are there? ____________

Circle one pair of like terms. ____________

List another pair of like terms. ____________

What is the constant term? ____________

Simplify Algebraic Expressions

pp. 179–180

Simplify each expression by combing like terms.

1. $4x + 3x = \square\, x$
2. $10 + 4y + 6y = 10 + \square\, y$
3. $15a + 6b - 3b + 2a = \square\, a + \square\, b$
4. $3t + 1 + 8t - 6 = \square\, t - \square$
5. $2m - 4k + 3 - 8m + 2 = \square\, m - \square\, k + \square$

Helping You Remember

Constant is a word used in everyday English as well as in mathematics. Write the definition of *constant*. Explain how the English definition can help you remember how *constant* is used in mathematics.

____

____

NAME ______________________ DATE __________ PERIOD __________

4-3 Solving Equations by Adding or Subtracting

What You'll Learn

Skim the Examples for Lesson 4-3. Write two things you already know about solving equations by adding or subtracting.

1. ____________________________________

____

2. ____________________________________

____

Active Vocabulary

New Vocabulary Write the correct term next to each definition.

____________________ ▶ undo each other

____________________ ▶ a mathematical sentence that contains an equals sign (=)

____________________ ▶ a value for the variable that makes an equation true

Vocabulary Link *Inverse operations* can have non-mathematical meanings as well. For each activity, name the inverse operation that would undo the activity.

turning on a light switch ____________________

driving 5 miles north ____________________

tying a shoelace ____________________

opening a window ____________________

Lesson 4-3

Main Idea	Details

Solve Equations by Adding
pp. 184–185

Fill in the blanks to solve each equation.

1. $x - 2 = 7$

$\square = \square$

$x = \square$

2. $y - 4 = -3$

$\square = \square$

$y = \square$

3. $b - 1\frac{2}{3} = \frac{1}{6}$

$\square = \square$

$b = \square$

4. $c - 6.2 = -9.7$

$\square = \square$

$c = \square$

Solve Equations by Subtracting
pp. 185–186

Model the following situation by drawing algebra tiles. Then solve.

Grace and Carrie have 14 necklaces combined. Carrie has 9 necklaces. How many does Grace have?

Helping You Remember

How is adding the same number of blocks to each side of a balance scale like the Addition Property of Equality?

____

____

____

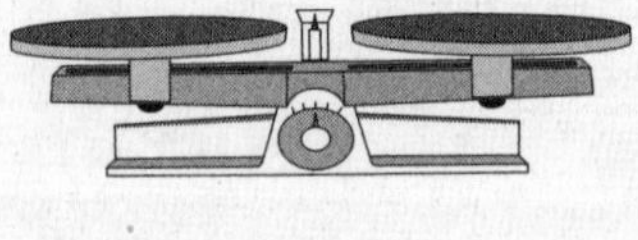

4-4 Solving Equations by Multiplying or Dividing

What You'll Learn

Skim the lesson. Write two things you already know about solving equations by multiplying and dividing.

1. ______________________________

____

2. ______________________________

____

____

Active Vocabulary

Review Vocabulary Label the diagram with the correct terms. *(Lesson 4-2)*

coefficient ▶

constant ▶

variable ▶

$$6x + 2$$

______ ______ ______

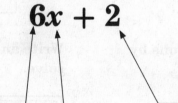

Vocabulary Link *Variable* and *constant* are two words used in everyday English. Find the definitions of *variable* and *constant* using a dictionary. List an example of something in your life that is variable. List an example of something in your life that is constant.

____

____

____

____

____

Lesson 4-4

NAME ____________________ DATE __________ PERIOD __________

Main Idea | **Details**

Solve Equations by Dividing
pp. 191–192

Fill in the blanks to solve each equation.

1. $3m = 18$

$$\frac{3m}{\square} = \frac{18}{\square}$$

$$m = \square$$

2. $-5n = 35$

$$\frac{-5n}{\square} = \frac{35}{\square}$$

$$n = \square$$

3. $0.6s = -42$

$$\frac{0.6s}{\square} = \frac{-42}{\square}$$

$$s = \square$$

4. $-8t = -48$

$$\frac{-8t}{\square} = \frac{-48}{\square}$$

$$t = \square$$

Solve Equations by Multiplying
p. 193

Write an equation to represent the model below. Then solve.

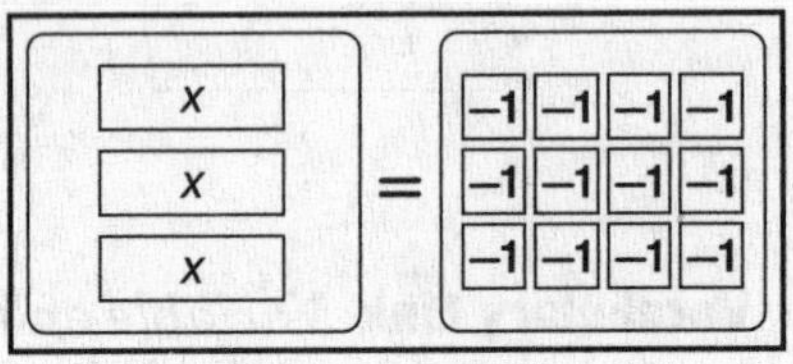

____

Helping You Remember

Write two examples of equations that can be solved using each of the four properties of equations below.

Addition Property of Equality: ______________________________

Subtraction Property of Equality: ______________________________

Multiplication Property of Equality: ______________________________

Division Property of Equality: ______________________________

NAME ______________________ DATE ____________ PERIOD ____________

4-5 Solving Two-Step Equations

What You'll Learn

Scan the text in Lesson 4-5. Write two facts you learned about solving two-step equations.

1. ______________________________

____

2. ______________________________

____

Active Vocabulary

Review Vocabulary Identify the following inverse operations. Draw a line from each operation to its inverse. *(Lessons 4-3 and 4-4).*

addition	addition
subtraction	subtraction
multiplication	multiplication
division	division

New Vocabulary Define the following terms from this lesson.

two-step equation ▶ ______________________________

____

Vocabulary Link *Two-step equations* can be illustrated by real-world examples. Consider the two-step process of putting on socks and putting on shoes. Explain how to "undo" the process. Write an example of another real-world process that takes two steps to "undo".

____

____

____

Lesson 4-5

Main Idea | **Details**

Solve Two-Step Equations
pp. 199–201

Justify each step used in solving the equation.

$6x - 14 = 16$

$6x - 14 + 14 = 16 + 14$ ______________________

$6x = 30$ ______________________

$\frac{6x}{6} = \frac{30}{6}$ ______________________

$x = 5$ ______________________

Solve the equation given the justification for each step.

$\frac{y}{9} + 4 = -2$

☐ = ☐ Subtraction Property of Equality

☐ = ☐ Simplify.

☐ = ☐ Multiplication Property of Equality

$y =$ ☐ Simplify.

Helping You Remember

List the steps you would use in the order you would use them to "undo" each equation.

$2x + 17 = 35$ ______________________

____

$\frac{x}{6} - 1 = 18$ ______________________

____

$\frac{x + 3}{4} = 5$ ______________________

____

NAME ______________________ DATE ____________ PERIOD ____________

4-6 Writing Equations

What You'll Learn

Scan the text under the *Now* heading. List two things you will learn about in the lesson.

1. ______________________

2. ______________________

Lesson 4-6

Active Vocabulary

Review Vocabulary Complete the table below listing words that indicate each operation. Use the words below the table. *(Lesson 1-1)*

Addition	Subtraction	Multiplication	Division

decreased by	difference
increased by	less
less than	more than
product	quotient
sum	times
total	twice

Main Idea	Details

Write Two-Step Equations
pp. 205–206

Answer each question using the information below.

Miguel and Carla spent $64 at the bookstore combined. Carla spent $15 less than Miguel.

1. Who spent less money at the bookstore? ____________
2. How much less? ____________
3. Write an expression to represent the amount of money Miguel spent, in terms of m. ____________
4. Write an expression to represent the amount of money Carla spent, in terms of m. ____________
5. Write an equation to represent the amount Miguel and Carla spent combined, in terms of m. ____________
6. How much did each person spend at the bookstore?

____

Two-Step Verbal Problems
p. 206

Write a verbal sentence to represent the equation below. Then solve.

$$\frac{x}{12} + 4 = 16$$

____

____

Helping You Remember

Write a word problem that can be solved using a two-step equation. Solve the equation.

____

____

____

NAME ______________________ DATE ____________ PERIOD ________

CHAPTER 4

Expressions and Equations

Tie It Together

Complete the graphic organizer to review writing and solving equations.

Georgia and her brother collected cans for a recycling program. Georgia collected three more than twice as many cans as her brother. They collected a total of 213 cans. How many cans did each person collect?

Steps		Example
Write the equation by translating from words to symbols, to the equation.	→	
Combine ___ terms.	→	
Work _________ through the order of operations to "undo" operations. ________________ _________ first.	→	
________________ next.	→	
Check your answer by ________________ ________.	→	
Answer the question that was asked.	→	

NAME ______________________ DATE __________ PERIOD __________

CHAPTER 4

Expressions and Equations

Before the Test

Review the ideas you listed in the table at the beginning of the chapter. Cross out any incorrect information in the first column. Then complete the table by filling in the third column.

K What I know...	W What I want to find out...	L What I learned...

Math Online Visit *glencoe.com* to access your textbook, more examples, self-check quizzes, personal tutors, and practice tests to help you study for concepts in Chapter 4.

Are You Ready for the Chapter Test?

Use this checklist to help you study.

☐ I used my Foldable to complete the review of all or most lessons.

☐ I completed the Chapter 4 Study Guide and Review in the textbook.

☐ I took the Chapter 4 Practice Test in the textbook.

☐ I used the online resources for additional review options.

☐ I reviewed my homework assignments and made corrections to incorrect problems.

☐ I reviewed all vocabulary from the chapter and their definitions.

Study Tips

- When you are preparing to read new material, scan the text first, briefly looking over headings, highlighted text, pictures, and call out boxes. Think of questions you might answer as you read.

NAME ______________________ DATE ____________ PERIOD ____________

CHAPTER 5

Multi-Step Equations and Inequalities

Before You Read

Before you read the chapter, think about what you know about solving multi-step equations and inequalities. List three things you already know about them in the first column. Then list three things you would like to learn about them in the second column.

K What I know...	W What I want to find out...

FOLDABLES® Study Organizer Construct the Foldable as directed at the beginning of this chapter.

Note Taking Tips

- **A visual (graph, diagram, picture, chart) can present information in a concise, easy-to-study format.**
  Clearly label your visuals and write captions when needed.
- **When you take notes, you may wish to use a highlighting marker to emphasize important concepts.**

NAME _______________ DATE _______ PERIOD _______

CHAPTER 5 Multi-Step Equations and Inequalities

Key Points

Scan the pages in the chapter and write at least one specific fact concerning each lesson. For example, in the lesson on inequalities, one fact might be that an inequality is a mathematical sentence that compares quantities that are not equal. After completing the chapter, you can use this table to review for your chapter test.

Lesson	Fact
5-1 Perimeter and Area	
5-2 Solving Equations with Variables on Each Side	
5-3 Inequalities	
5-4 Solving Inequalities	
5-5 Solving Multi-Step Equations and Inequalities	

5-1 Perimeter and Area

What You'll Learn

Scan the text in Lesson 5-1. Write two facts you learned about perimeter and area as you scanned the text.

1. ____
____

2. ____
____

Active Vocabulary

Review Vocabulary Fill in each blank with the correct term or phrase. *(Lesson 1-2)*

variable ▶ a ____________ or ____________ that represents an ____________ value

algebraic expression ▶ an ____________ with a least one ____________ and one ____________

New Vocabulary Match the term with its definition by drawing a line to connect the two.

area	distance around a geometric figure
perimeter	equation that shows a relationship among certain quantities
formula	measure of the surface enclosed by a figure

Main Idea	Details
Perimeter pp. 221–222	**Complete the diagram by labeling each figure so that its perimeter is equal to 24 millimeters.**

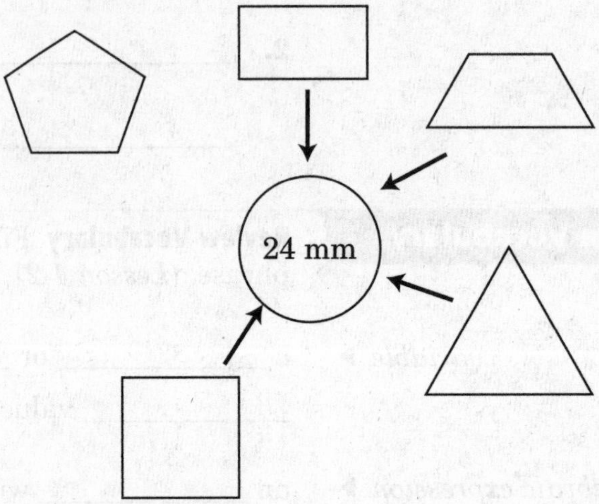

Area pp. 222–223	**Model a triangle with height of 12 inches and base length of 10 inches. Then find its area.**

Helping You Remember

Compare and contrast the units used for perimeter and area. Explain why area uses square units and perimeter does not.

____

____

____

NAME _______________ DATE __________ PERIOD __________

5-2 Solving Equations with Variables on Each Side

What You'll Learn

Scan the text under the *Now* heading. List two things you will learn about in the lesson.

1. ______________________________

____

2. ______________________________

____

Active Vocabulary

Review Vocabulary Fill in each blank with the correct term or phrase. *(Lessons 4-2 and 4-3)*

like terms ▶ terms that contain the same ______________

simplest form ▶ an algebraic expression that has no ________ terms and no ______________

simplifying the expression ▶ You can use the ____________ ____________ to combine like terms.

Additive or Subtraction Properties of Equality ▶ allows you to ________ or ________ the same quantity from each side of an ____________ to keep the two sides equal

Vocabulary Link *Equality* is a word that is used in everyday English. Find the definition of *equality* by using a dictionary. List an example of something in your life that has *equality*.

____

____

____

____

Lesson 5-2

Main Idea	Details
Equations with Variables on Each Side pp. 229–230	**Model the equation $3x - 2 = 5x - 4$ using algebra tiles. Then solve.**

Fill in the blanks to solve each equation.

1. $2x + 5 = 3x$

____ $= x$

2. $7b + 5 = -3b - 10$

$10b =$ ____

$b =$ ____

3. $21 - 16t = 4t - 14$

____ $t = -35$

$t =$ ____

4. $0.8y + 1.6 = 0.6y - 1$

$0.2y =$ ____

$y =$ ____

5. $9a - 3 = 15$

$9a =$ ____

$a =$ ____

6. $18x + 6 = 9 - 3x$

____ $x = 3$

$x =$

Helping You Remember

Write an equation with a variable on both sides, along with all the steps needed to solve the equation. Trade with a partner. Then each of you should explain verbally why each step in solving the equation was carried out.

____

____

NAME ______________________ DATE __________ PERIOD __________

5-3 Inequalities

What You'll Learn

Skim the lesson. Write two things you already know about inequalities.

1. ____________________

2. ____________________

Active Vocabulary

Review Vocabulary Write the term next to each definition. *(Lesson 2-1)*

__________ ▶ a number less than zero

__________ ▶ the counting numbers, their opposites, and zero

__________ ▶ a number greater than zero

New Vocabulary Write the definition next to the term.

inequality ▶ ____________________

Vocabulary Link *Inequality* is a word that is used in everyday English. Find the definition of *inequality* using a dictionary. Explain how the English definition can help you remember how *inequality* is used in mathematics.

____

Lesson 5-3

Main Idea	Details

Write Inequalities
pp. 234–235

Fill in the organizer with words that describe the symbols.

Inequality Symbols

>	<	≥	≤
________	________	________	________
________	________	________	________
________	________	________	________
________	________	________	________

Graph Inequalities
p. 236

Write an inequality for each model.

1. ____________________

2. ____________________

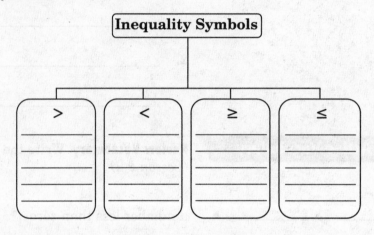

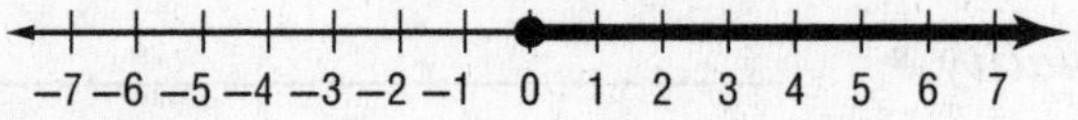

Helping You Remember

Write a paragraph explaining how to graph an inequality to a classmate that was absent from class the day it was taught. Include an explanation of the symbols used, as well as the use of open and closed dots or points.

____

____

____

____

NAME ______________________________ DATE ____________ PERIOD ____________

5-4 Solving Inequalities

What You'll Learn

Scan Lesson 5-4. List two headings you would use to make an outline of this lesson.

1. ____
____

2. ____
____

Active Vocabulary

Review Vocabulary Use the diagram to fill in each blank with the correct term. *(Lesson 5-1)*

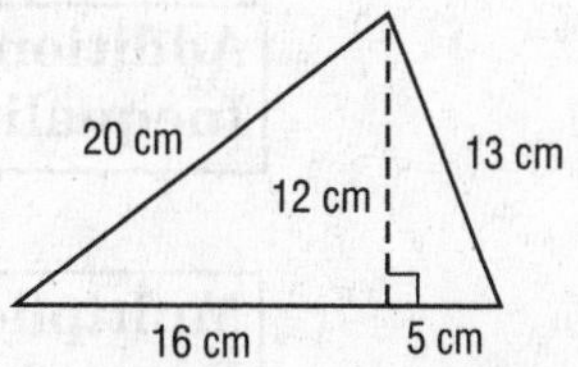

formula ▶

perimeter ▶

area ▶

The ________ for the triangle is 126 cm^2 and the ________ for the triangle is 54 cm.

$A = \frac{1}{2}bh$ is the ____________ for the area of a triangle.

Vocabulary Link *Addition* and *Subtraction Properties* allow you to add or subtract the same quantities to each side of an equation or inequality with the sentence remaining true. List an example of something in everyday life that you have to do "the same on both sides" to keep it equal or the same.

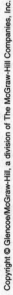

____

____

____

Main Idea	Details

Solve Inequalities by Adding or Subtracting
pp. 241–242

Draw an arrow and match the correct property needed to solve the inequality. Then solve each inequality.

Solve Inequalities by Multiplying or Dividing by a Positive Number
pp. 242–243

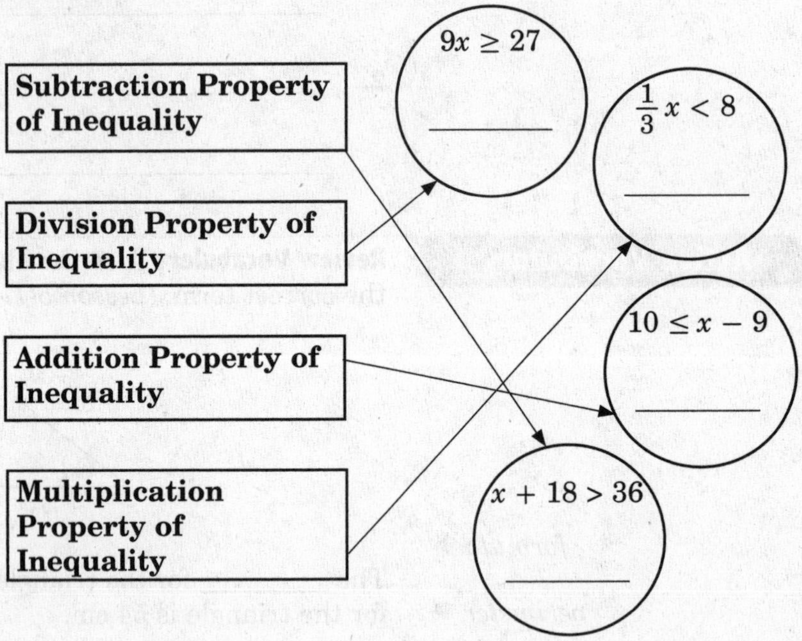
Subtraction Property of Inequality
Division Property of Inequality
Addition Property of Inequality
Multiplication Property of Inequality
$9x \geq 27$
$\frac{1}{3}x < 8$
$10 \leq x - 9$
$x + 18 > 36$

Multiply or Divide an Inequality by a Negative Number
pp. 243–244

Model the solution of the inequality on the number line.

$-4y \geq 24$

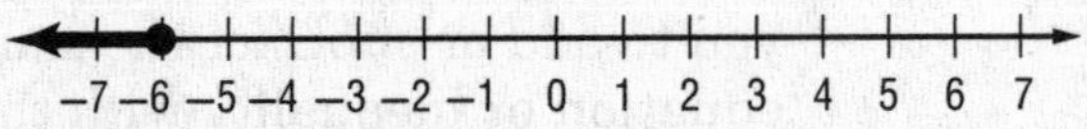

Helping You Remember

In the boxes below, write examples of inequalities in which the sign does and does not reverse. Write at least three examples in each box.

NAME ______________________ DATE __________ PERIOD __________

5-5 Solving Multi-Step Equations and Inequalities

What You'll Learn

Skim the Examples for Lesson 5-5. Predict two things you think you will learn about solving multi-step equations and inequalities.

1. ____

____

2. ____

____

Active Vocabulary

Review Vocabulary Fill in each blank with the correct term or phrase. *(Lesson 4-1)*

Distributive Property ▶ To __________ a sum or difference by a number, __________ each term inside the __________ by the number outside of the __________.

New Vocabulary Write the definition next to each term.

null or empty set ▶ ____

____

identity ▶ ____

____

Vocabulary Link *Null* is a word that is used in everyday English. Find the definition of *null* using a dictionary. Explain how the English definition can help you remember how *null* is used in mathematics.

____

____

____

Main Idea	Details

Solve Equations and Inequalities with Grouping Symbols
pp. 248–249

Complete the organizer by following the steps given to solve the inequality.

Steps in Solving Multi-Step Equations and Inequalities $\quad 2(x - 3) \quad 4(x + 3) - 6x$

Step 1: Use the Distributive Property to remove parentheses → **Step 1:**

↓

Step 2: Combine like terms on same side → **Step 2:**

↓

Step 3: Use the Addition or Subtraction Properties → **Step 3:**

↓

Step 4: Use the Multiplication or Division Properties → **Step 4:**

No Solution or All Numbers as Solutions
p. 250

Write an equation that has a solution that is an identity. Then write an equation with a null or empty set.

identity:

null or empty set:

Helping You Remember

An identity is an equation that is true for every value of the variable. A null or empty set occurs when an equation has no solution. Write and solve an example of each type of equation.

____

____

NAME ______________________ DATE ____________ PERIOD ________

CHAPTER 5 Multi-Step Equations and Inequalities

Tie It Together

Complete the graphic organizer to compare and contrast equations and inequalities.

	Multi-Step Equations	Inequalities
Symbol(s)	______________________ ______________________ ______________________	______________________ ______________________ ______________________
Steps to solve	______________________ ______________________ ______________________ ______________________ ______________________	______________________ ______________________ ______________________ ______________________ ______________________
Graph	______________________ ______________________ ______________________ ______________________	______________________ ______________________ ______________________ ______________________

NAME ______________________ DATE __________ PERIOD __________

CHAPTER 5

Multi-Step Equations and Inequalities

Before the Test

Review the ideas you listed in the table at the beginning of the chapter. Cross out any incorrect information in the first column. Then complete the table by filling in the third column.

K What I know...	W What I want to find out...	L What I learned...

Math Online Visit *glencoe.com* to access your textbook, more examples, self-check quizzes, personal tutors, and practice tests to help you study for concepts in Chapter 5.

Are You Ready for the Chapter Test?

Use this checklist to help you study.

☐ I used my Foldable to complete the review of all or most lessons.

☐ I completed the Chapter 5 Study Guide and Review in the textbook.

☐ I took the Chapter 5 Practice Test in the textbook.

☐ I used the online resources for additional review options.

☐ I reviewed my homework assignments and made corrections to incorrect problems.

☐ I reviewed all vocabulary from the chapter and their definitions.

Study Tips

- On test day, look over the entire test to get an idea of its length and scope so that you can pace yourself. Answer what you know first, then go back and complete the problems you skipped. When finished, check for errors. Don't change an answer unless you are certain you are correct.

NAME ______________________________ DATE ____________ PERIOD ____________

CHAPTER 6 Ratio, Proportion, and Similar Figures

Before You Read

Before you read the chapter, respond to these statements.

1. Write an **A** if you agree with the statement.
2. Write a **D** if you disagree with the statement.

Before You Read	Ratio, Proportion, and Similar Figures
	• A ratio is a comparison of quantities by addition.
	• Unit rates are useful when comparing prices.
	• Ratios are used to change a measurement from one unit to another.
	• When two figures are proportional, they will have the same side and angle measures.
	• A scale drawing is sometimes proportional to the actual object.

FOLDABLES® Study Organizer Construct the Foldable as directed at the beginning of this chapter.

Note Taking Tips

- **To help you organize data, create study cards when taking notes, recording and defining vocabulary words, and explaining concepts.**
- **When taking notes, use a table to make comparisons about the new material.** Determine what will be compared, decide what standards will be used, and then use what is known to find similarities and differences.

NAME ______________________________ DATE ____________ PERIOD ____________

CHAPTER 6

Ratio, Proportion, and Similar Figures

Key Points

Scan the pages in the chapter and write at least one specific fact concerning each lesson. For example, in the lesson on solving proportions, one fact might be that the cross products of any proportion are equal. After completing the chapter, you can use this table to review for your chapter test.

Lesson	Fact
6-1 Ratios	
6-2 Unit Rates	
6-3 Converting Rates and Measurements	
6-4 Proportional and Nonproportional Relationships	
6-5 Solving Proportions	
6-6 Scale Drawings and Models	
6-7 Similar Figures	
6-8 Dilations	
6-9 Indirect Measurement	

NAME ________________________ DATE ____________ PERIOD ____________

6-1 Ratios

What You'll Learn

Scan Lesson 6-1. List two headings you would use to make an outline of this lesson.

1. ______________________________

____

2. ______________________________

____

Active Vocabulary

Review Vocabulary Define *simplify* in your own words. *(Lesson 1-3)*

simplify ▶ ______________________________

____

New Vocabulary Fill in each blank with the correct term or phrase.

ratio ▶ a ____________ of two ____________ by division that is usually written in __________ form

Vocabulary Link *Ratio* is a word that is used in everyday English. Find the definition of *ratio* using a dictionary. List two examples of real-life ratios.

____

____

____

____

Lesson 6-1

Main Idea	Details

Write Ratios as Fractions in Simplest Form

pp. 265–266

Cross out the ratio that is not equivalent to the following ratio.

16 girls out of 24 students

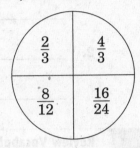

$\frac{2}{3}$	$\frac{4}{3}$
$\frac{8}{12}$	$\frac{16}{24}$

Simplify Ratios Involving Measurements

p. 266

Write each ratio in simplest form.

1. 15 cans out of 9 cases = ☐

2. 4 rings to 7 bracelets = ☐

3. 2 c to 32 oz = ☐

4. 16 in. to 4 ft = ☐

5. 11 dramas out of 17 DVDs = ☐

6. 6 hours 14 days = ☐

Helping You Remember

Ratios can represent part to part, part to whole, or whole to part relationships. Write a problem that can be expressed with these three ratios. Include the ratios in your description.

____

____

____

____

NAME ______________________ DATE ____________ PERIOD ____________

6-2 Unit Rates

What You'll Learn

Skim the Examples for Lesson 6-2. Predict two things you think you will learn about unit rates.

1. ______________________________

____

2. ______________________________

____

Active Vocabulary

Review Vocabulary Fill in each blank with the correct term or phrase. *(Lesson 5-1)*

area ▶ The ________ of the ________ enclosed by a figure.

New Vocabulary Write the definition next to each term.

rate ▶ ______________________________

____

unit rate ▶ ______________________________

____

Vocabulary Link *Rate* is a word that is used in everyday English. Find the definition of *rate* using a dictionary. Write two examples of rates used in everyday life.

____

____

____

Lesson 6-2

Main Idea	Details

Find Unit Rates
p. 270

Complete the Venn diagram by writing the phrases in the correct position. Use the phrases below the diagram.

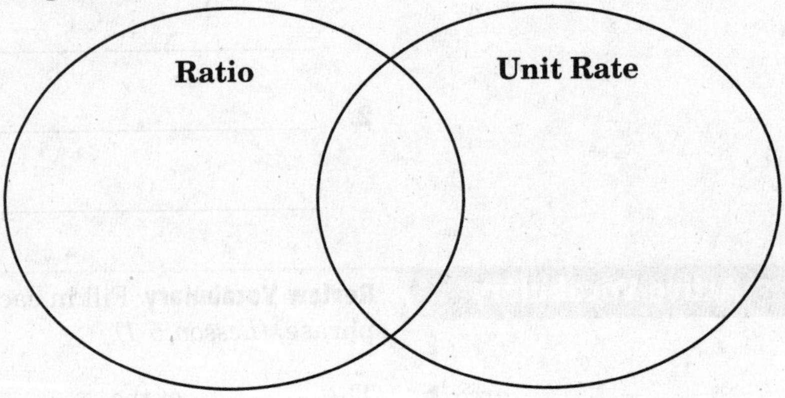

usually a fraction	same units
4 miles to 1,000 feet	different units
has 1 in denominator	uses "out of"
5 inches per second	uses "per"

Compare Unit Rates
p. 271

Fill in each blank with <, >, or = to compare the unit rates.

1. 10 notebooks for $12 ☐ 15 notebooks for $18.75
2. 12 cans for $4.20 ☐ 20 cans for $6
3. 171 miles with 9 gallons ☐ 300 miles with 15 gallons
4. 4,000 ft in 16 seconds ☐ 7,500 ft in 30 seconds
5. 77 pages in 1 hour ☐ 108 pages in 120 minutes

Helping You Remember

The word rate is part of the term *unit rate*.

Explain how a rate can be written as a unit rate.

____

____

____

6-3 Converting Rates and Measurements

What You'll Learn

Skim Lesson 6-3. Predict two things that you expect to learn based on the headings.

1. ______________________________

____

2. ______________________________

____

Active Vocabulary

Review Vocabulary Write the correct term next to each definition. *(Lesson 4-3)*

______________ ▶ a mathematical sentence that contains an equals sign, (=), showing that two expressions are equal

______________ ▶ a value for the variable that makes an equation true

______________ ▶ operations that "undo" each other

New Vocabulary Fill in each blank with the correct term or phrase.

dimensional analysis ▶ the process of including __________ of __________ as factors when you compute

Vocabulary Link *Analysis* is a word that is used in everyday English. Find the definition of *analysis* using a dictionary. Explain how the English definition can help you remember how *analysis* is used in mathematics.

____

____

____

____

Lesson 6-3

Main Idea | **Details**

Dimensional Analysis
pp. 275–276

Fill in each conversion factor and solve each problem.

1. Convert 8 cups of juice per 1 gallon of water to cups of juice per quart of water.

$$\frac{8 \text{ cups}}{1 \text{ gallon}} \cdot \frac{1 \text{ gallon}}{\square \text{ quarts}} = \boxed{}$$

2. Convert 110 millimeters per meter to millimeters per centimeter.

$$\frac{110 \text{ mm}}{1 \text{ m}} \cdot \frac{1 \text{ m}}{\square} = \boxed{}$$

3. Convert 80 ounces per minute to ounces per second.

$$\frac{80 \text{ oz}}{1 \text{ min}} \cdot \frac{\square}{\square} = \boxed{}$$

Convert Between Systems
pp. 276–277

Fill in the diagram to complete the steps to convert between measurement systems. Use the terms *simplify, ratio, divide, and conversion factor*.

Step 1	Step 2	Step 3
Set up a ______ of the measurement you are converting. →	Write a ______ using the units you are converting to. →	______ out common units and then ______.

Helping You Remember

Describe how to convert 18 pounds to kilograms. In the conversation factor, which quantity is in the numerator and which quantity is in the denominator?

____

____

____

NAME ______________________ DATE __________ PERIOD __________

6-4 Proportional and Nonproportional Relationships

What You'll Learn

Scan the text under the *Now* heading. List two things you will learn about in the lesson.

1. ______________________________

____

____

2. ______________________________

____

Active Vocabulary

Review Vocabulary Match the term with the correct definition by drawing a line to connect the two. *(Lessons 6-1 and 6-2)*

ratio	a simplified rate with a denominator of 1
unit rate	a ratio with two quantities that have different kinds of units
rate	comparison of two quantities by division

New Vocabulary Write the correct term next to each definition.

______________________ ▶ the relationship between two quantities where the ratio or rate is *not* constant

____

______________________ ▶ a constant ratio or unit rate of a proportion

______________________ ▶ a relationship where two quantities have a constant ratio or rate

Lesson 6-4

Lesson 6-4 *(continued)*

Main Idea | **Details**

Identify Proportions
pp. 281–282

Fill in the organizer about proportions.

What are proportions?	How can proportions be written with numbers?
Examples	**Nonexamples**

(Center: **Proportions**)

Describe Proportional Relationships
p. 282

Fill in the blanks so that each table represents a proportional relationship.

1.

cups of juice	2	4	6	8
cups of water	6			

2.

oranges	1		3	
apples	4	8		16

3.

cats	3	9		
dogs	5		25	35

Helping You Remember

All relationships can be classified as either proportional or nonproportional. Think of six examples of numerical relationships that you use on a daily basis and classify each as proportional or nonproportional.

____

____

____

____

NAME ______________________ DATE ____________ PERIOD ____________

6-5 Solving Proportions

What You'll Learn

Scan the text in Lesson 6-5. Write two facts you learned about solving proportions.

1. ____

____

2. ____

____

____

Active Vocabulary

Review Vocabulary Write the definition next to each term. *(Lesson 4-3)*

equation ▶ ____

____

____

solution ▶ ____

____

New Vocabulary Fill in each blank with the correct word or phrase.

proportion ▶ an ____________ that states that two ____________ or rates are ____________

cross products ▶ If $\frac{a}{b} = \frac{c}{d}$, then ________ = ________.

Lesson 6-5

Main Idea	Details

Proportions
pp. 287–288

Complete the organizer for cross products.

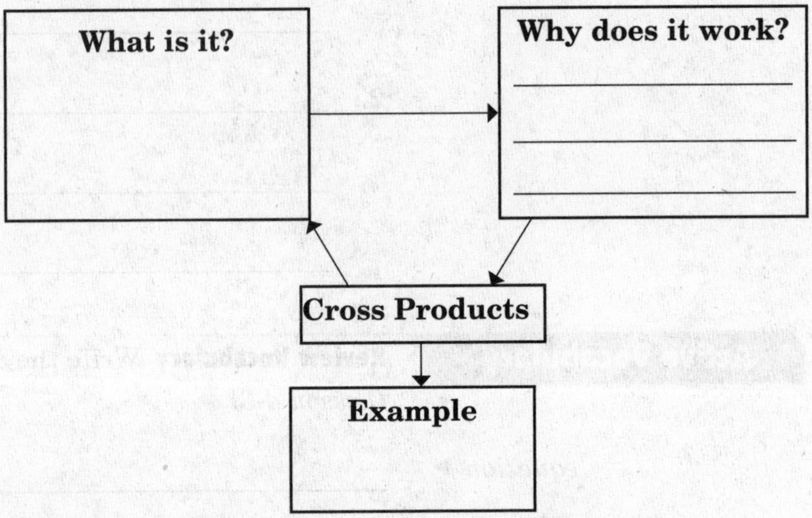

What is it?
Why does it work?
Cross Products
Example

Use Proportions to Solve Problems
pp. 288–289

Fill in each blank with a ratio that forms a proportion.

1. $\frac{4}{12} = \square$

2. $\frac{10}{20} = \square$

3. $\frac{3.0}{1.8} = \square$

4. $\frac{7}{28} = \square$

Helping You Remember

Proportion is a common word in the English language. Use a dictionary to look up its definition. Explain how the definition from the dictionary can help you remember the mathematical definition of *proportion*.

____

____

____

NAME ______________________ DATE ____________ PERIOD ____________

6-6 Scale Drawings and Models

Lesson 6-6

What You'll Learn

Skim the lesson. Write two things you already know about scale drawings and models.

1. ____

____

____

2. ____

____

Active Vocabulary

New Vocabulary Match the term with the correct definition by drawing a line to connect the two.

scale factor	a ratio of a given length on a scale model or drawing to its corresponding length on the actual object
scale model	a diagram used to represent an object that is too large or small to be drawn at actual size
scale	the relationship between the measurements on a drawing or model and the measurements of the real object
scale drawing	a model used to represent an object that is too large or small to be built at actual size

Vocabulary Link *Scale* is a word that is used in everyday English. Find the definition of *scale* using a dictionary. Explain how the English definition can help you remember how *scale* is used in mathematics.

____

____

____

Main Idea | **Details**

Use Scale Drawings and Models
pp. 294–295

Fill in the table using the information provided.

The actual measurements of a 5-room apartment are in the table below. Use the scale of $\frac{1}{2}$ in. = 4 ft to find the missing lengths of the drawing.

Room	Living room	Kitchen	Bathroom	Bedroom #1	Bedroom #2
Actual length (ft)	14	10	6	8	12
Drawing length (in.)					

Construct Scale Drawings
p. 296

Construct a scale drawing of the floor plan of the 5-room apartment above using the values you calculated for the lengths of each room.

Helping You Remember

Explain how a scale is different than a scale factor.

____

____

____

NAME ______________________ DATE ____________ PERIOD ____________

6-7 Similar Figures

What You'll Learn

Scan Lesson 6-7. List two headings you would use to make an outline of this lesson.

1. ______________________________

____

2. ______________________________

____

Active Vocabulary

Review Vocabulary Write the definition of *proportion* in your own words. *(Lesson 6-4)*

proportion ▶ ______________________________

____

New Vocabulary Quadrilateral *DEFG* ~ quadrilateral *HIJK*. Label the diagram with the correct terms. Use each term once.

similar figures ▶

corresponding parts ▶

congruent ▶

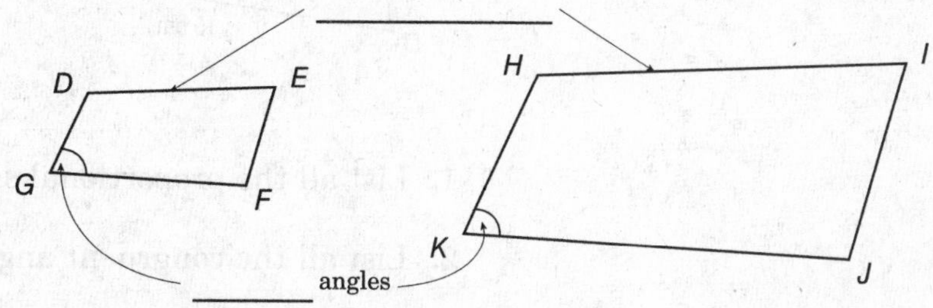

DEFG and *HIJK* are ______________.

Vocabulary Link *Similar* and *congruent* are two words used in everyday English. Find the definitions of *similar* and *congruent* using a dictionary.

____

____

____

Lesson 6-7

Main Idea	Details
Corresponding Parts of Similar Figures pp. 301–302	**Use the triangles below.**

List the congruent angles.	**List the corresponding sides.**
______________ ______________	______________ ______________

Scale Factors
p. 303

Fill in each blank to answer the questions about the figures below.

ABCD ~ *EFGH*

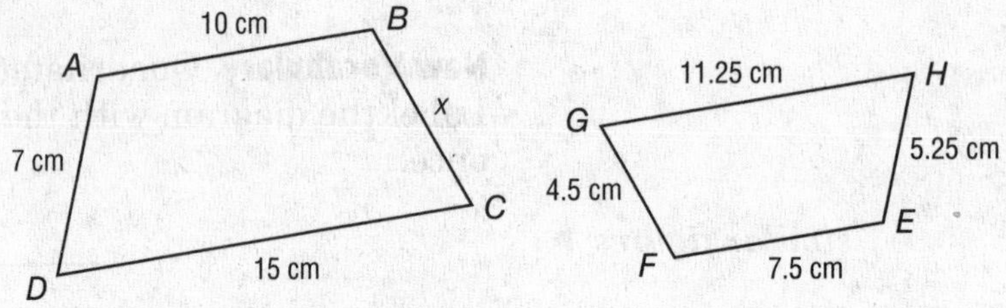

1. List all the proportional sides. ______________

2. List all the congruent angles.

____

3. What is the scale factor? ________

4. What is the value of *x*? ________

Helping You Remember

Make a list of what you learned about similar figures.

____

____

NAME ______________________ DATE __________ PERIOD __________

6-8 Dilations

What You'll Learn

Skim the Examples for Lesson 6-8. Predict two things you think you will learn about dilations.

1. ______________________________

____

2. ______________________________

____

Active Vocabulary

Review Vocabulary Match the term with its definition by drawing a line to connect the two. *(Lessons 1-4 and 2-7)*

x-coordinate	a pair of numbers used to locate any point on a coordinate plane
ordered pair	second number in an ordered pair
coordinate plane	movement of a geometric figure
y-coordinate	first number in an ordered pair
transformation	formed by the intersection of two number lines that meet at right angles at their zero points

New Vocabulary Fill in each blank with the correct term or phrase.

dilation ▶ a ______________ that enlarges or reduces a figure by a ______ factor

Lesson 6-8

Main Idea	Details
Dilations pp. 307–309	**Compare and contrast the three types of transformations by completing the diagram below, using the terms under the diagram.**

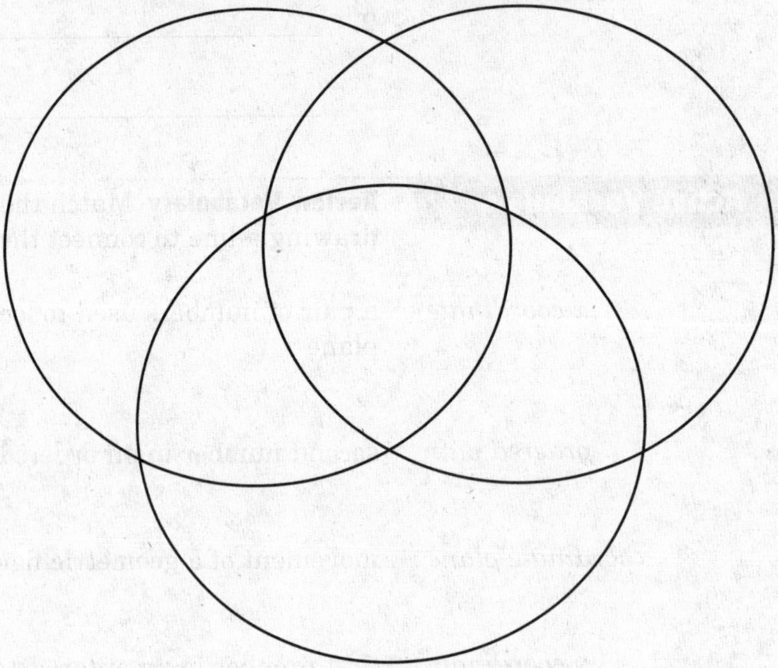

slide
enlarge or reduce
keeps original shape
same orientation
changes size (unless k = 1)
same size
changes orientation
results in similar figures

Helping You Remember

Dilation is a word used in everyday English as well as in mathematics. Write the definition of *dilation*. Explain how the English definition can help you remember how *dilation* is used in mathematics.

____

____

NAME ______________________ DATE __________ PERIOD __________

6-9 Indirect Measurement

What You'll Learn

Skim the Examples for Lesson 6-9. Predict two things that you will learn about indirect measurement.

1. ______________________________

____

2. ______________________________

____

Active Vocabulary

Review Vocabulary Write the correct term next to each definition. *(Lessons 6-5 and 6-7)*

__________ ▶ a statement of equality of two or more ratios

__________ ▶ If $\frac{a}{b} = \frac{c}{d}$, then $ad = cb$.

__________ ▶ figures that have the same shape but not necessarily the same size

New Vocabulary Fill in each blank with the correct term or phrase.

indirect measurement ▶ allows you to use the properties of __________ to find measurements that are difficult to measure __________

Vocabulary Link *Indirect* is a word that is used in everyday English. Find the definition of *indirect* using a dictionary. Explain how the English definition can help you remember how *indirect* is used in mathematics.

____

____

____

Main Idea	Details

Indirect Measurement
p. 313

Model the following situation with a labeled drawing. Then solve.

A flagpole casts a shadow that is 32 feet long. At the same time, a statue that is 7 feet tall casts a shadow that is $17\frac{1}{2}$ feet long. How tall is the flagpole?

Surveying Methods
p. 314

Fill in the blank of the missing measure.

The triangles below are similar. What is the distance from Springdale to Porter?

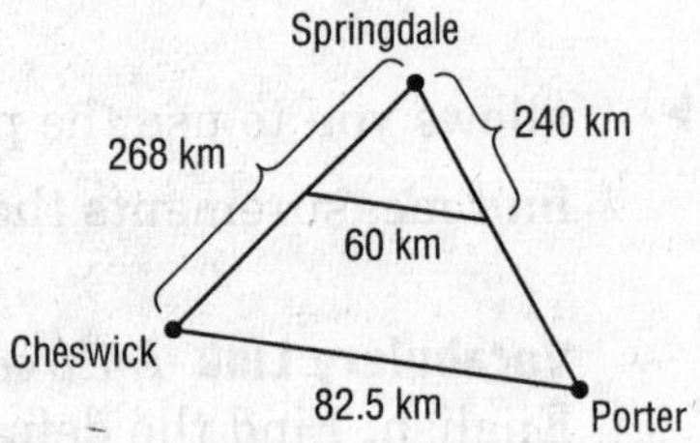

Helping You Remember

Write a paragraph explaining how to find a missing measurement using similar triangles to a classmate that was absent from class the day it was taught. Include an example.

____

____

____

NAME ______________________________ DATE ____________ PERIOD ____________

CHAPTER 6

Ratio, Proportion, and Similar Figures

Tie It Together

Describe how ratios are used in each application.

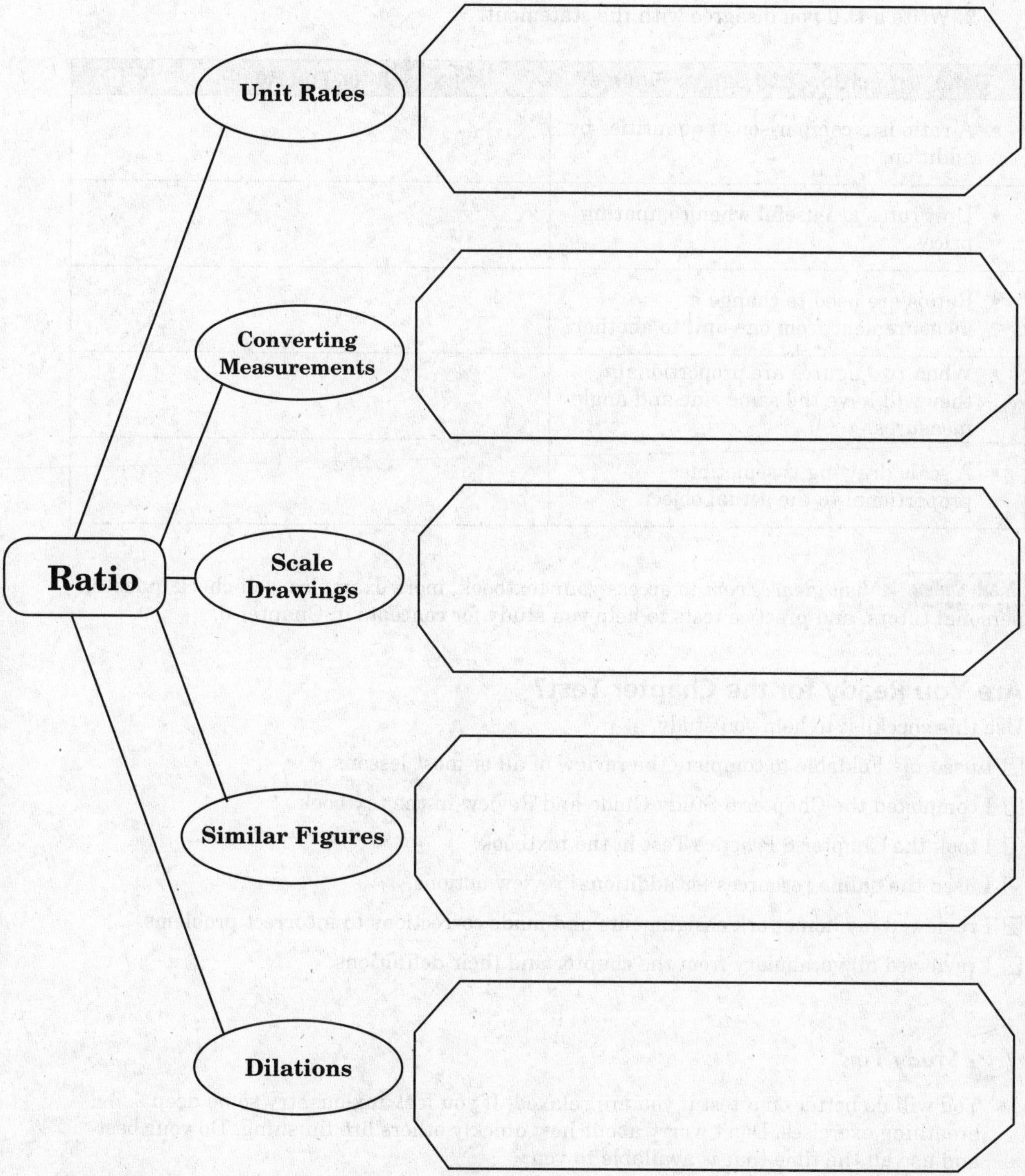

NAME _______________ DATE _______ PERIOD _______

CHAPTER 6

Ratio, Proportion, and Similar Figures

Before the Test

Now that you have read and worked through the chapter, think about what you have learned and complete the table below. Compare your previous answers with these.

1. Write an **A** if you agree with the statement.

2. Write a **D** if you disagree with the statement.

Ratio, Proportion, and Similar Figures	After You Read
• A ratio is a comparison of quantities by addition.	
• Unit rates are useful when comparing prices.	
• Ratios are used to change a measurement from one unit to another.	
• When two figures are proportional, they will have the same side and angle measures.	
• A scale drawing is sometimes proportional to the actual object.	

Math Online Visit *glencoe.com* to access your textbook, more examples, self-check quizzes, personal tutors, and practice tests to help you study for concepts in Chapter 6.

Are You Ready for the Chapter Test?

Use this checklist to help you study.

☐ I used my Foldable to complete the review of all or most lessons.

☐ I completed the Chapter 6 Study Guide and Review in the textbook.

☐ I took the Chapter 6 Practice Test in the textbook.

☐ I used the online resources for additional review options.

☐ I reviewed my homework assignments and made corrections to incorrect problems.

☐ I reviewed all vocabulary from the chapter and their definitions.

Study Tips

- You will do better on a test if you are relaxed. If you feel anxious, try some deep breathing exercises. Don't worry about how quickly others are finishing. Do your best and use all the time that is available to you.

NAME ______________________ DATE __________ PERIOD __________

CHAPTER 7 Percent

Before You Read

Before you read the chapter, respond to these statements.

1. Write an **A** if you agree with the statement.
2. Write a **D** if you disagree with the statement.

Before You Read	Percent
	• A percent is a comparison of a number and 100.
	• To write a decimal as a percent, divide by 100.
	• Percents can be written in fraction, decimal, or percent form.
	• A percent proportion is solved with cross products.
	• The percent equation can only be used with percents in their fraction form.

FOLDABLES® Study Organizer Construct the Foldable as directed at the beginning of this chapter.

Note Taking Tips

- **When taking notes, write clean and concise explanations.** Someone who is unfamiliar with the math concepts should be able to read your explanations and learn from them.
- **If your instructor points out definitions or procedures from your text, write a reference page in your notes.** You can then write these referenced items in their proper place in your notes after class.

NAME ______________________ DATE ____________ PERIOD ____________

CHAPTER 7 Percent

Key Points

Scan the pages in the chapter and write at least one specific fact concerning each lesson. For example, in the lesson on simple and compound interest, one fact might be that the formula used to solve simple interest problems is $I = prt$. After completing the chapter, you can use this table to review for your chapter test.

Lesson	Fact
7-1 Fractions and Percents	
7-2 Fractions, Decimals, and Percents	
7-3 Using the Percent Proportion	
7-4 Find Percent of a Number Mentally	
7-5 Using Percent Equations	
7-6 Percent of Change	
7-7 Simple and Compound Interest	
7-8 Circle Graphs	

7-1 Fractions and Percents

What You'll Learn

Scan Lesson 7-1. List two headings you would use to make an outline of this lesson.

1. ______________________________

____

2. ______________________________

____

Active Vocabulary

Review Vocabulary Write the correct term next to each definition. *(Lessons 6-1 and 6-4)*

______________ ▶ a comparison of two quantities by division that is often written in fraction form

______________ ▶ describes the relationship between two quantities where the ratio or rate is not constant

____
______________ ▶ the k in the equation $y = kz$

______________ ▶ a relationship where two quantities have a constant ratio or rate

New Vocabulary Write the definition next to the term.

percent ▶ ______________________________

____

Vocabulary Link *Percent* is a word that is used in everyday English. Find the definition of *percent* using a dictionary. List two examples of everyday uses of *percents*.

____

____

Lesson 7-1

Main Idea | **Details**

Percents as Fractions
pp. 331–332

Fill in the fraction that completes the circle. Then define the relationship between the four parts.

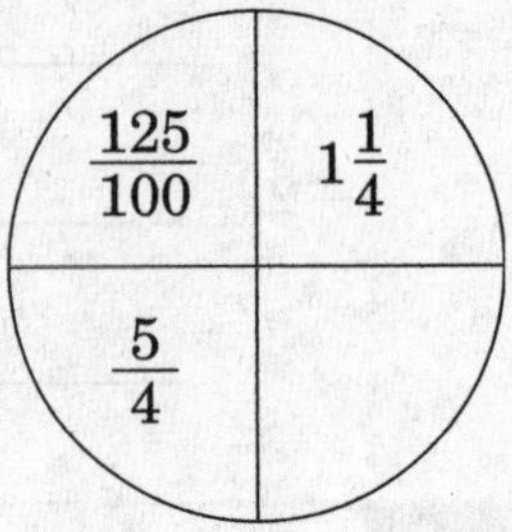

The relationship:

Fractions as Percents
pp. 332–333

Fill in each blank of the proportion to find the percent following the given steps.

What percent is 16 out of 24?

$\frac{16}{24} = \square$ Write a proportion.

$\square \cdot \square = \square \cdot \square$ Cross products are equal.

$\square = \square$ Multiply.

$\square = n$ Simplify.

So, $\frac{16}{24} = \square$ or $\square$

Helping You Remember

Write and solve two questions where you would use proportions and percents to solve.

____

____

____

NAME _______________ DATE _______ PERIOD _______

7-2 Fractions, Decimals, and Percents

What You'll Learn

Skim the lesson. Write two things you already know about fractions, decimals, and percents.

1. ______________________________

____

2. ______________________________

____

Active Vocabulary

Review Vocabulary Fill in each blank with the correct term or phrase. *(Lessons 6-1, 6-5, and 7-1)*

ratio ▶ a __________ of two quantities by ________ that is usually written in ________ form

percent ▶ a _____ that compares a number to ____

proportion ▶ an __________ that states that two __________ or rates are __________

cross products ▶ If $\frac{a}{b} = \frac{c}{d}$, then ________ = ________.

Vocabulary Link The historical form of *percent* was *per cent.* Use a dictionary to look up the words *per* and *cent.* Relate these two meanings to the current definition of *percent.*

____

____

____

____

Lesson 7-2

NAME ______________________ DATE ____________ PERIOD ____________

Lesson 7-2 *(continued)*

Main Idea	Details

Percents and Decimals
pp. 337–339

Complete the diagram by filling out each box with a description and example of each process described.

% → fraction

% → decimal

fraction → %

decimal → %

Compare Fractions, Decimals, and Percents
p. 339

At a local school, 22% of students walk to school, 0.35 take the bus, and three eighths are driven in a car. The rest of the students ride their bikes. Which of these groups are the largest?

____

____

____

Helping You Remember

Suppose you are a percent. Write a paragraph describing your experiences as a percent.

____

____

____

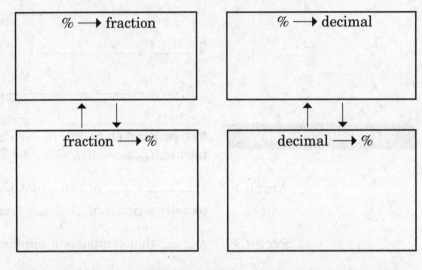

NAME ______________________ DATE __________ PERIOD __________

7-3 Using the Percent Proportion

What You'll Learn

Skim the Examples in Lesson 7-3. Predict two things that you think you will learn about using percent proportions.

1. ____

____

2. ____

____

Active Vocabulary

Review Vocabulary Match the term with its definition by drawing a line connecting the two. *(Lessons 4-3, 6-1, and 6-5)*

equation	a ratio that compares a number to 100
cross products	a mathematical sentence that contains an equals sign, (=), showing that two expressions are equal
proportion	an equation that states that two ratios or rates are equal
percent	If $\frac{a}{b} = \frac{c}{d}$, then $ad = cb$.

New Vocabulary Write the definition next to the term.

percent proportion ▶ ____

____

____

Vocabulary Link *Percent* and *proportion* are words that are used in everyday English. Find the definition of *percent* and *proportion* using a dictionary. How can their individual definitions help you remember what a percent proportion is?

____

____

____

____

Lesson 7-3

NAME ______________________ DATE ____________ PERIOD ____________

Main Idea	**Details**

The Percent Proportion
pp. 345–347

Complete the model for the *percent proportion*.

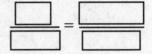

$$\frac{\square}{\square} = \frac{\square}{\square}$$

Complete the organizer. Write the types of percent problems. Then write a word problem to show an example for each.

Type of Percent Problems

- Find the ______.
- Find the ______.
- Find the ______.

Helping You Remember

Fill in each blank to identify the whole, the part, and the percent in the following percent proportion.

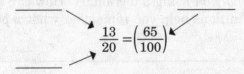

______ → $\frac{13}{20} = \left(\frac{65}{100}\right)$ ← ______

______ ↗

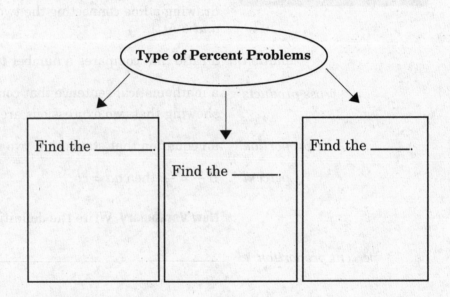

NAME ______________________ DATE __________ PERIOD __________

7-4 Find Percent of a Number Mentally

What You'll Learn

Skim Lesson 7-4. Predict two things you expect to learn based on the headings and the Key Concept box.

1. ______________________________
____

2. ______________________________
____

Active Vocabulary

Review Vocabulary Write the correct term next to each definition. *(Lessons 6-2, 6-5, 7-1, and 7-2)*

______________ ▶ a ratio that compares a number to 100

______________ ▶ two equivalent ratios, one with a denominator of 100 and the other that compares the part to the whole.

______________ ▶ a ratio of two quantities that have different kinds of units

______________ ▶ an equation that states two ratios or rates are equal

Vocabulary Link *Mental* and *math* are both words used in everyday English. Look up *mental* and *math* in the dictionary. Explain how the two words fit together to be the *mental math* that is used in everyday mathematics.

____

____

____

____

Lesson 7-4

Main Idea	Details

Find Percent of a Number Mentally
pp. 351–352

Complete the organizer with two ways that you can mentally find 40% of $700.

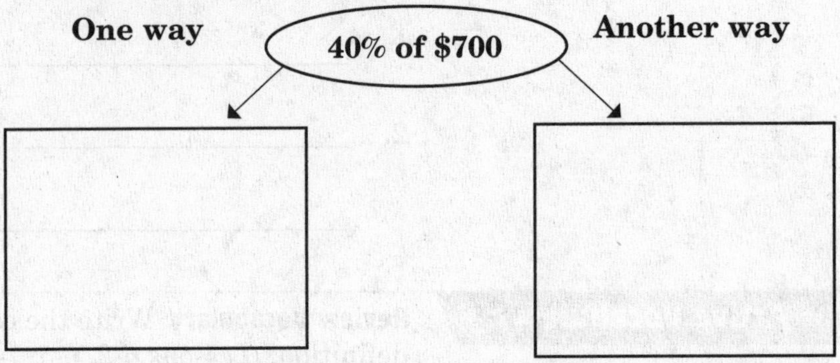

Estimates with Percents
pp. 352–353

Fill in the table with the mental strategy of how you found the estimate. Use a different strategy each time.

Estimate the Answer	Describe your Strategy
150% of 98	
76% of 160	
$\frac{1}{2}$% of 280	

Helping You Remember

There are situations when an exact answer is needed. There are other times when an estimate is good enough. Give examples of when an exact answer and an estimate are appropriate. Explain your reasoning.

____

____

____

____

NAME ______________________ DATE __________ PERIOD __________

7-5 Using Percent Equations

What You'll Learn

Scan the text in Lesson 7-5. Write two facts you learned about solving using percent equations as you scanned the text.

1. ______________________________

____

2. ______________________________

____

Active Vocabulary

New Vocabulary Write the definition next to each term. *(Lessons 4-3, 6-5, and 7-3)*

proportion ▶ ______________________________

____

percent proportion ▶ ______________________________

____

____

equation ▶ ______________________________

____

cross products ▶ ______________________________

New Vocabulary Fill in each blank with the correct term or phrase.

percent equation ▶ an __________ form of the ______________ in which the percent is written as a ________

Main Idea	Details

Percent Equations
pp. 357–359

Complete the organizer. Write the type of percent problem using the terms part, whole, and percent. Then solve using the percent equation.

Percent Equation
part = percent × whole

82.5 is what percent of 550?	**What is 15% of 142?**	**14 is 20% of what number?**
Find the ______.	Find the ______.	Find the ______.
____________	____________	____________
______	______	______

Fill in each blank using the information below.

Terri and Kraig each bought an MP3 player. Terri paid $45 minus an 18% discount. Kraig has a coupon for 15% off, which is a $6 discount.

1. What is the original price of Kraig's player? ☐
2. How much did Terri pay after her discount? ☐
3. Suppose Kraig pays 6.5% and Terri pays 6% sales tax. How much did Kraig and Terri spend total for both players, including tax? Round to the nearest cent if needed. ☐

Helping You Remember

The label over each oval represents what is missing from a percent equation. In each oval, write and solve a percent equation to find that missing information.

Missing Whole	**Missing Part**	**Missing Percent**
____________	____________	____________
____________	____________	____________
______	______	______

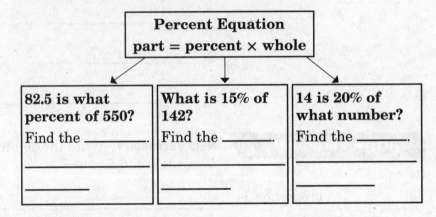

7-6 Percent of Change

What You'll Learn

Skim Lesson 7-6. Predict two things that you expect to learn based on the headings and Key Concept box.

1. ____

____

2. ____

____

Active Vocabulary

New Vocabulary Match the term with the correct definition by drawing a line to connect the two.

percent of decrease	ratio that compares the change in quantity to the original amount
selling price	the amount the price of an item is increased above the price the store paid for an item
discount	a positive percent of change
percent increase	total amount consumer pays for item
markup	a negative percent of change
percent of change	the amount by which the regular price of an item is reduced

Vocabulary Link *Percent of change* is a term that is used in everyday English. List two ways in which *percent of change* is used in everyday life.

____

____

____

Main Idea	Details

Find Percent of Change
pp. 364–365

Find the percent of change. Round to the nearest tenth if necessary. Then fill in each blank with either *percent increase* or *percent decrease*.

1. from 89° to 77° ________ percent of __________.

2. from \$45 to \$58 ________ percent of __________.

3. In her first week of training, Cherie ran 5.4 miles. Six weeks later she ran 7.1 miles. What is the percent change in the 6 weeks? ________

4. Marcus weighed 113 pounds at the start of wrestling season and 106 pounds at the end. What is the percent change from the start to the end of the season?

   ____

Using Markup and Discount
pp. 365–366

Complete the Venn diagram for the terms *markup* and *discount*. Use the terms *percent increase, percent decrease, percent of change, positive, and negative*.

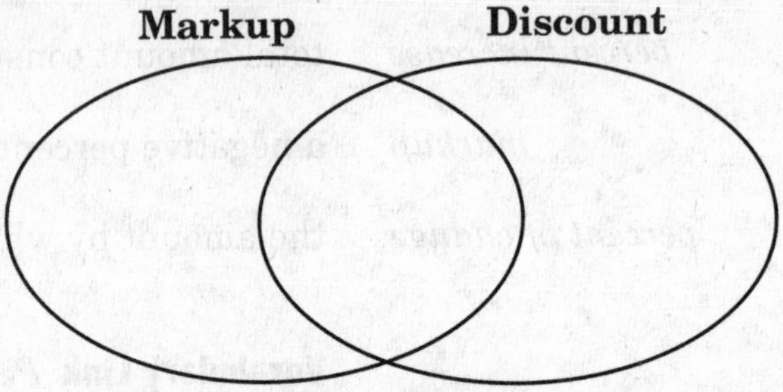

Helping You Remember For a percent of increase, is the percent of change always positive or negative? Why? For a percent of decrease, is the percent of change always positive or negative? Why?

____

____

____

____

NAME ______________________________ DATE ____________ PERIOD ____________

7-7 Simple and Compound Interest

What You'll Learn

Scan the text under the *Now* heading. List two things you will learn about in this lesson.

1. ____

____

2. ____

____

Active Vocabulary

principal ▶

interest ▶

annual interest rate ▶

time year ▶

New Vocabulary Label the diagram with the correct terms.

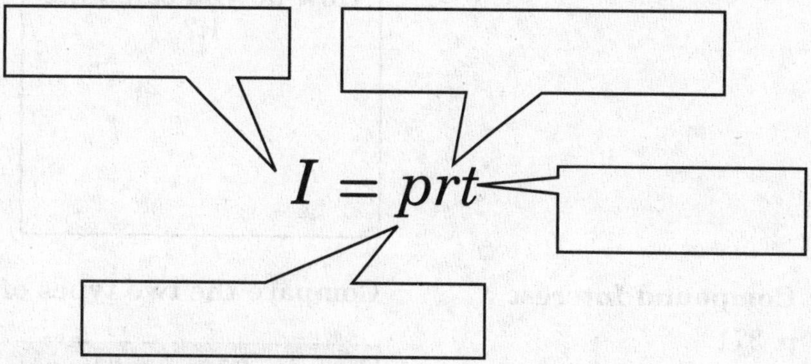

Vocabulary Link *Principal* is a word used in everyday English as well as in mathematics. Write the definition of *principal*. Explain how the English definition can help you remember how *principal* is used in mathematics.

____

____

____

____

Lesson 7-7

Main Idea	Details

Simple Interest
pp. 370–371

Complete the organizer. Sample answers are given.

What is it?

Simple Interest

How do you calculate it?

Write and solve a simple interest problem.

Compound Interest
p. 371

Compare the two types of interest.

Types of Interest	Description
Simple	
Compound	

Helping You Remember

Write a real-world problem dealing with compound interest. Find the answer then trade with a partner and solve.

____

____

____

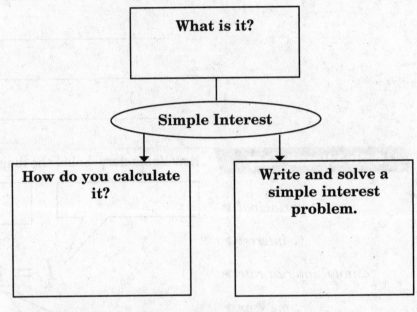

NAME ______________________ DATE __________ PERIOD __________

7-8 Circle Graphs

What You'll Learn

Skim the Examples for Lesson 7-8. Predict two things you think you will learn about circle graphs.

1. ______________________

____

2. ______________________

____

Active Vocabulary

Review Vocabulary Write the term next to each definition. *(Lessons 3-1 and 7-1)*

______________ ▶ a decimal whose digits end

______________ ▶ a decimal whose digits repeat in groups of one or more without end

______________ ▶ a ratio that compares a number to 100

New Vocabulary Write the definition next to the term.

circle graph ▶ ______________________

____

Vocabulary Link A circle graph displays data. Look up *data* in a dictionary. Find a sentence in the lesson that uses that word.

Definition: ______________________

Sentence: ______________________

____

____

Lesson 7-8

NAME ______________________ DATE __________ PERIOD __________

Main Idea | **Details**

Circle Graphs
pp. 376–377

Model a circle graph by following the steps below.
Count the number of students in your classroom. Then count the number of students that are wearing various colored shirts. For example, there may be 7 students wearing blue shirts, 3 wearing pink shirts, and 6 wearing white shirts. Construct and label a circle graph with your data.

Analyze Circle Graphs
p. 378

Answer each question using the circle graph.

250 students were surveyed about their favorite activities. The results are in the circle graph.

How many students favor computer? ___

How many more students favor sports than favor swimming? ___

Which activity is most favored?

______

Favorite Activity

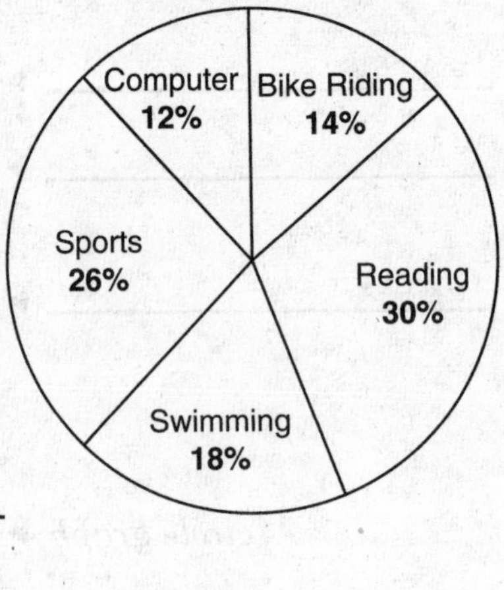

Helping You Remember

Describe how to construct a circle graph in detail to a classmate who was absent.

____

____

____

____

____

NAME ______________________ DATE ____________ PERIOD ____________

CHAPTER 7

Percent

Tie It Together

Use the fraction $\frac{3}{4}$ to show how to convert to different forms of numbers.

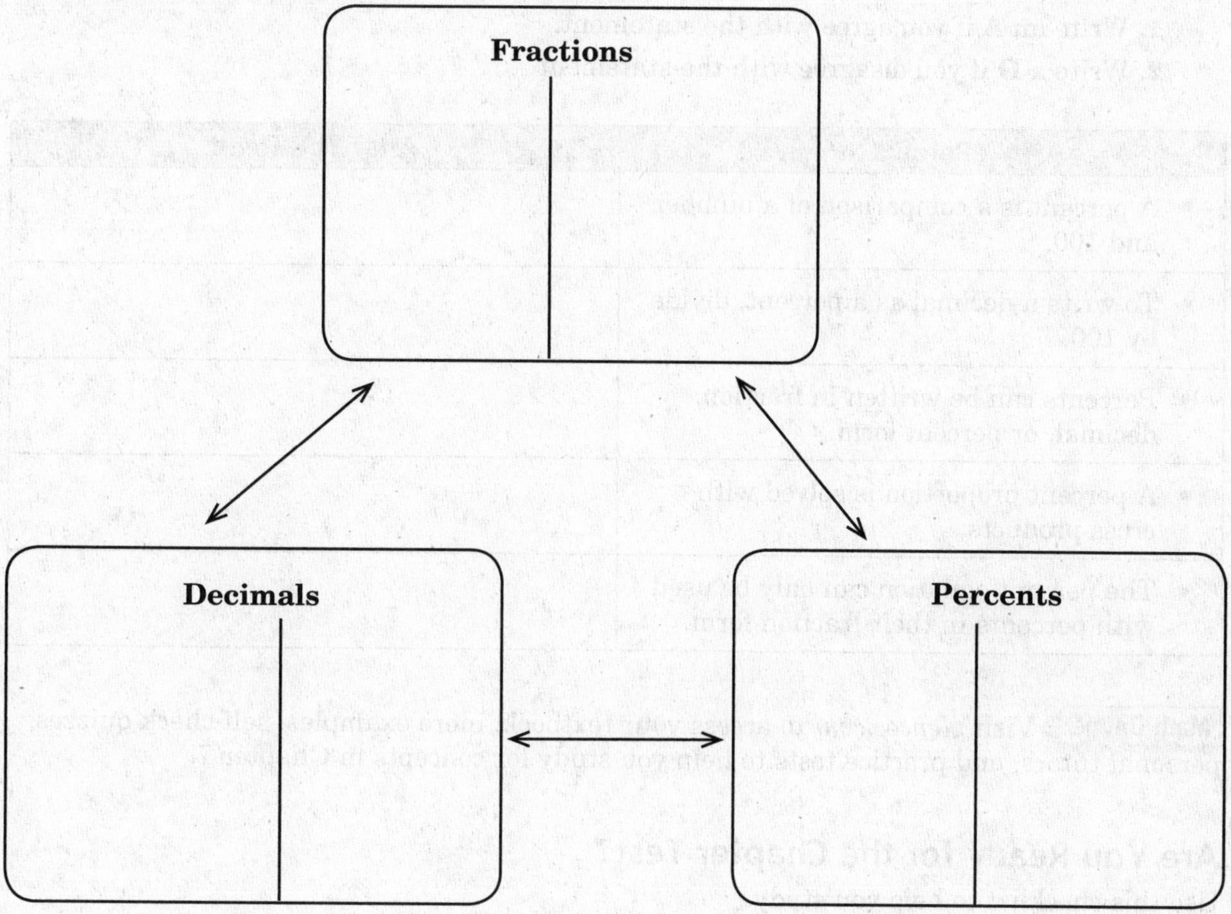

Complete the proportion with the words "part", "whole", and the symbols % and 100.

NAME ______________________ DATE __________ PERIOD __________

CHAPTER 7 Percent

Before the Test

Now that you have read and worked through the chapter, think about what you have learned and complete the table below. Compare your previous answers with these.

1. Write an **A** if you agree with the statement.
2. Write a **D** if you disagree with the statement.

Percent	After You Read
• A percent is a comparison of a number and 100.	
• To write a decimal as a percent, divide by 100.	
• Percents can be written in fraction, decimal, or percent form.	
• A percent proportion is solved with cross products.	
• The percent equation can only be used with percents in their fraction form.	

Math Online Visit *glencoe.com* to access your textbook, more examples, self-check quizzes, personal tutors, and practice tests to help you study for concepts in Chapter 7.

Are You Ready for the Chapter Test?

Use this checklist to help you study.

☐ I used my Foldable to complete the review of all or most lessons.

☐ I completed the Chapter 7 Study Guide and Review in the textbook.

☐ I took the Chapter 7 Practice Test in the textbook.

☐ I used the online resources for additional review options.

☐ I reviewed my homework assignments and made corrections to incorrect problems.

☐ I reviewed all vocabulary from the chapter and their definitions.

Study Tips

- Designate a place to study at home that is free of clutter and distraction. Try to study at about the same time each afternoon or evening so that it is part of your routine.

NAME _______________ DATE _______________ PERIOD _______________

CHAPTER 8 Linear Functions and Graphing

Before You Read

Before you read the chapter, respond to these statements.

1. Write an **A** if you agree with the statement.
2. Write a **D** if you disagree with the statement.

Before You Read	Linear Function and Graphing
	• In a function, a member of the domain can be paired with more than one member of the range.
	• An arithmetic sequence has a common ratio between each term.
	• A linear function has both straight and curved lines.
	• If a rate of change is proportional, its graph will be a straight line.
	• In the equation $y = 5x + 3$, the slope is 3.

FOLDABLES® Study Organizer Construct the Foldable as directed at the beginning of this chapter.

Note Taking Tips

- **When you take notes, write concise definitions in your own words.** Add examples that illustrate the concepts.
- **When taking notes, write down a question mark by anything you do not understand.** Before your next quiz, ask your instructor to explain these sections.

NAME ______________________ DATE ____________ PERIOD ________

CHAPTER 8

Linear Functions and Graphing

Key Points

Scan the pages in the chapter and write at least one specific fact concerning each lesson. For example, in the lesson on slope, one fact might be that positive slopes represent a rate of increase. After completing the chapter, you can use this table to review for your chapter test.

Lesson	Fact
8-1 Functions	
8-2 Sequences and Equations	
8-3 Representing Linear Functions	
8-4 Rate of Change	
8-5 Constant Rate of Change and Direct Variation	
8-6 Slope	
8-7 Slope-Intercept Form	
8-8 Writing Linear Equations	
8-9 Prediction Equations	
8-10 Systems of Equations	

NAME ______________________ DATE __________ PERIOD __________

8-1 Functions

What You'll Learn

Skim the Examples for Lesson 8-1. Predict two things you think you will learn about functions.

1. ______________________________

____

2. ______________________________

____

Active Vocabulary

New Vocabulary Match each term with its definition by drawing a line to connect the two.

function notation	a value that is chosen and does not depend on the other variable
independent variable	a value that depends on the input value
vertical line test	a way to write an equation using $f(x)$
dependent variable	use to determine if a graph is a function

Vocabulary Link *Independent* and *dependent* are two words used in everyday English. Find the definitions of *independent* and *dependent* using a dictionary. Write an example of a variable in everyday life that is *independent.* Write an example of a variable that is *dependent*.

____

____

____

____

____

Lesson 8-1

Main Idea	Details

Relations and Functions
pp. 395–396

Complete the organizer for *functions.*

What is a relation?	**When is a relation a function?**
Examples of Functions	**Nonexamples of Functions**

(center oval: **Relations and Functions**)

Function Notation
p. 396

Write the equation in function notation. Label *both* forms with the terms *independent variable* and *dependent variable.*

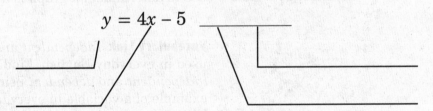

Equation **Function Notation**

$y = 4x - 5$ ____________

____________ ____________

____________ ____________

Describe Relationships
p. 397

In the function above, what is the value of y if $x = 7$? ___

Helping You Remember

Explain how the vertical line test works to tell whether a relation is a function.

____

____

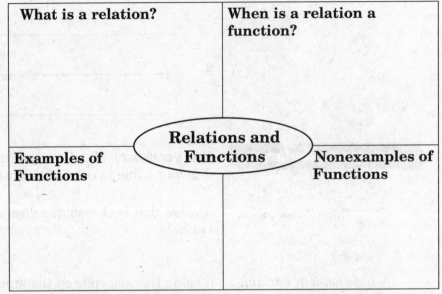

NAME ______________________ DATE __________ PERIOD __________

8-2 Sequences and Equations

What You'll Learn

Scan Lesson 8-2. List two headings you would use to make an outline of this lesson.

1. ______________________________________

____

2. ______________________________________

____

Active Vocabulary

New Vocabulary Label the diagram with the correct terms.

common difference ▶

arithmetic sequence ▶

term ▶

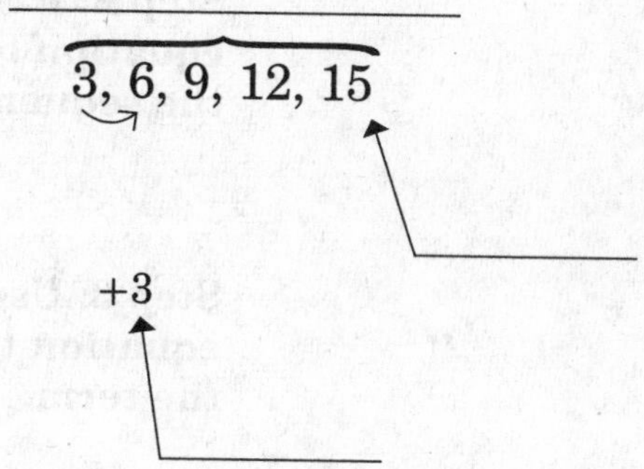

Main Idea

Details

Describe Sequences
p. 401

Fill in each blank to complete the arithmetic sequence.

Term Number (n)	1		3	
Term (t)	12		24	

The difference of the term numbers is __.

The common difference of the terms is __.

The equation that describes the sequence is ________.

Lesson 8-2

Main Idea	Details

Finding Terms
p. 402

Complete the organizer to find a term in an arithmetic sequence.

To find a term in an arithmetic sequence...	**14, 21, 28, 35, 42, 49, ... Find the 11th term.**
Step 1: Find the common difference. →	**Step 1:**
Step 2: Write an equation to describe the sequence. →	**Step 2:**
Step 3: Use the equation to find the term. →	**Step 3:**

Helping You Remember

Suppose you are an arithmetic sequence. Write a paragraph describing your experiences.

____

____

____

____

____

NAME ______________________ DATE __________ PERIOD __________

8-3 Representing Linear Functions

What You'll Learn

Scan the text in Lesson 8-3. Write two facts you learned about representing linear functions as you scanned the text.

1. ______________________________

____

2. ______________________________

____

____

Active Vocabulary

Review Vocabulary Write the term next to each definition. (*Lesson 1-5*)

______________ ▶ a mathematical sentence stating that two quantities are equal

______________ ▶ a relation where each member of the domain is paired with exactly one member in the range

New Vocabulary Fill in each blank with the correct term or phrase.

linear equation ▶ an equation whose graph is a ______

x-intercept ▶ the ______________ of the point at which the graph crosses the ________

y-intercept ▶ the ______________ of the point at which the graph crosses the ________

Lesson 8-3

Main Idea	Details

Solve Linear Equations
pp. 406–407

Fill in the blanks to complete each table. Write the ordered pairs under the table.

1. $y = 3x + 1$

x	y
−2	
	1
	4
2	

____

2. $y = -x + 2$

x	y
−1	
	2
2	0
3	−1

____

Graph Linear Equations
p. 408

Compare the two methods of graphing a linear function.

One Way:		Another Way:
Find the _______ by finding x when y = ____.	**Step 1**	Rewrite the _______ by solving for y.
Find the _______ by finding y when x = ____.	**Step 2**	Choose 4 values for x and find the _______ values for y.
Graph the _______ and connect them with a ____.	**Step 3**	Graph the _______ and connect them with a ____.

NAME ______________________ DATE ____________ PERIOD ____________

8-4 Rate of Change

What You'll Learn

Skim Lesson 8-4. Predict two things that you expect to learn based on the headings and the Key Concept Box.

1. ______________________________

____

2. ______________________________

____

Active Vocabulary

New Vocabulary Write the definition next to the term.

rate of change ▶ ______________________________

____

Main Idea

Details

Rate of Change
pp. 412–414

Model a graph with a *positive* and then a *negative* rate of change. Describe your graph with words.

Positive Rate of Change

Time and Distance Traveled

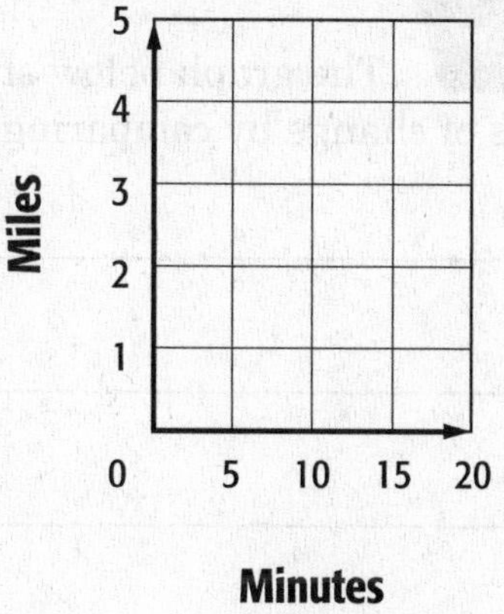

____

____

____

____

Negative Rate of Change

Time and Distance Traveled

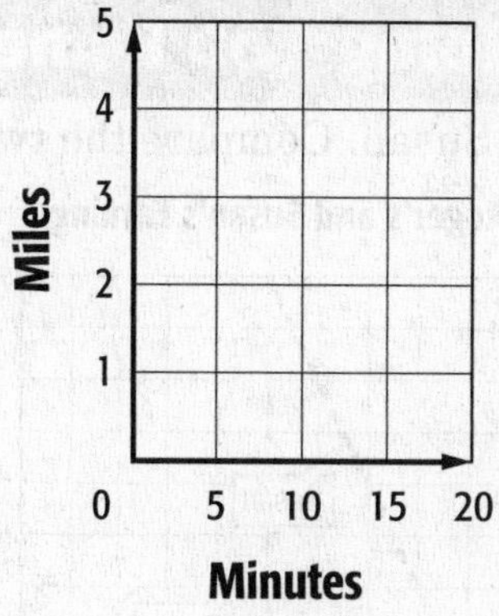

____

____

____

____

Main Idea	Details

Summarize the following situation.

Emily is filling a bathtub with water. She turns the faucet on, and 7 minutes later when the bathtub is full, she turns the faucet off.

Describe the rate of change. How would the graph of the water flow appear?

____

____

____

Describe another situation where the rate of change of the graph would appear the same.

____

____

____

Helping You Remember

The graph below shows the earnings of Roger and Susan. Compare the two rates of change by comparing the steepness of the lines.

Roger's and Susan's Earnings

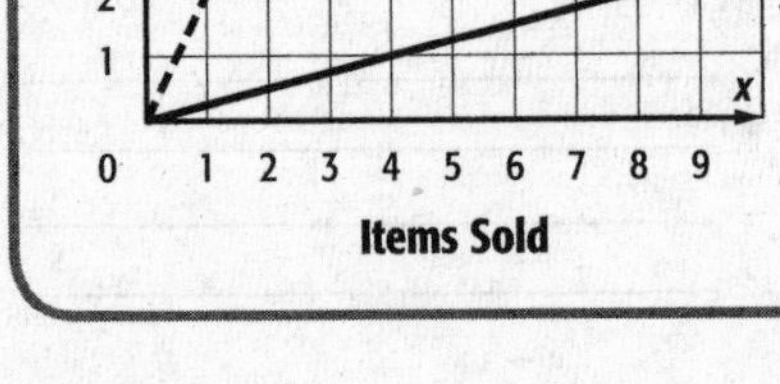

____

____

____

____

____

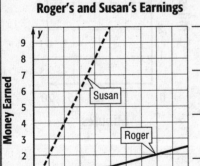

NAME ______________________ DATE __________ PERIOD __________

8-5 Constant Rate of Change and Direct Variation

What You'll Learn

Scan the text under the *Now* heading. List two things you will learn about in the lesson.

1. ______________________

____

2. ______________________

____

Active Vocabulary

Review Vocabulary Write the definition next to each term. *(Lessons 6-4 and 8-4)*

rate of change ▶ ______________________

____

proportion ▶ ______________________

New Vocabulary Write the correct term next to each definition.

______________ ▶ the constant of proportionality, the k in the equation $y = kx$

______________ ▶ the relationship between two quantities that results in a straight-line graph

______________ ▶ when the ratio between two variable quantities is constant

______________ ▶ a linear relationship where the rate of change between any two data points is the same

Lesson 8-5

Main Idea | **Details**

Constant Rate of Change
pp. 418–420

Cross out the set of coordinates in the circle that do not belong. Then describe the relationship.

(4, 14), (2, 7)	(8, 2), (12, 4)
(5, 3), (15, 9)	(1, 1), (6, 6)

The relationship is ______________________.

Direct Variation
pp. 420–421

Fill in the organizer about *direct variation*.

What is it?	**How can it be written using symbols?**
Direct Variation	
Examples	**Nonexamples**

Helping You Remember

All linear relationships are not proportional.

Give an example of a linear relationship that has a constant rate of change but is not proportional.

____

____

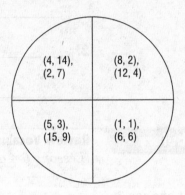

NAME ______________________ DATE __________ PERIOD __________

8-6 Slope

Lesson 8-6

What You'll Learn

Skim the lesson. Write two things you already know about slope.

1. ____

____

2. ____

____

Active Vocabulary

Review Vocabulary Fill in each blank with the correct term or phrase. *(Lessons 6-1 and 8-5)*

ratio ▶ a ____________ of two ____________ by division that is often written in ________ form

constant rate of change ▶ a ________________ where the rate of change between any two data points is ________

New Vocabulary Write the definition next to the term.

slope ▶ ____

____

Main Idea

Details

Slope
pp. 427–428

Match the different types of slopes to the correct coordinates by drawing a line to connect the two.

positive slope	$A(2, 4)$, $B(2, 5)$
negative slope	$C(-5, 3)$, $D(-3, 2)$
undefined slope	$E(7, 4)$, $F(-7, 4)$
zero slope	$G(-1, -3)$, $H(-3, -5)$

Main Idea | **Details**

Slope and Constant Rate of Change

pp. 428–429

Complete the organizer to find the slope of a line. Fill in each blank to find the slope in the example.

Find the slope of a line.

Step 1: Label the coordinates as ( ) and ( ).

Step 2: Substitute the coordinates into the ______________.

Step 3: Simplify.

Find the slope of line.

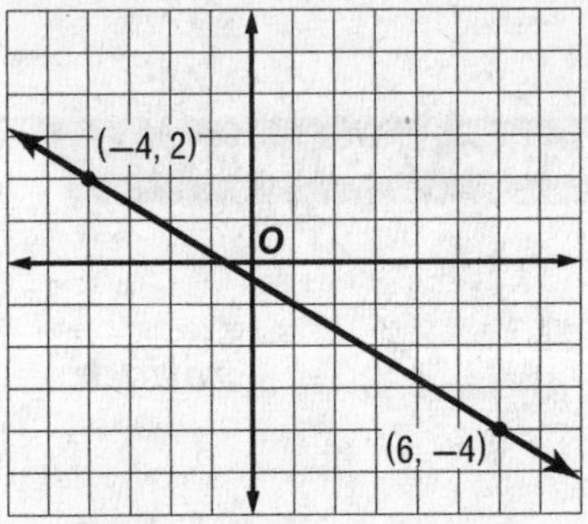

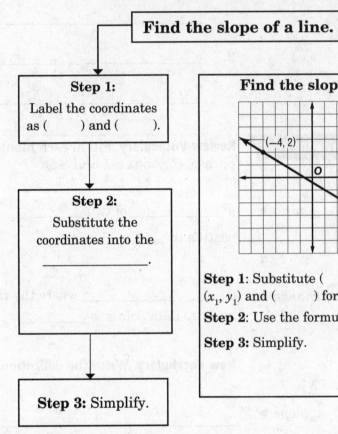

Step 1: Substitute ( ) for (x_1, y_1) and ( ) for (x_2, y_2).

Step 2: Use the formula.

Step 3: Simplify.

Helping You Remember

Use words to describe how a line appears with the given slopes.

positive: ______________________

negative: ______________________

zero: ______________________

undefined: ______________________

NAME ______________________ DATE __________ PERIOD __________

8-7 Slope-Intercept Form

What You'll Learn

Skim the Example for Lesson 8-7. Predict two things you think you will learn about slope-intercept form.

1. ______________________________

____

2. ______________________________

____

Active Vocabulary

Review Vocabulary Write the correct term next to each definition. *(Lessons 6-5 and 8-3)*

______________ ▶ a statement of equality of two or more ratios

______________ ▶ an equation whose graph is a straight line

______________ ▶ the first number of an ordered pair

______________ ▶ the second number of an ordered pair

New Vocabulary Write the definition next to the term.

slope-intercept form ▶ ______________________________

____

Lesson 8-7

Main Idea	Details

Find Slope and *y*-intercept
pp. 433–434

Identify the slope and *y*-intercept in each equation.

1. $y = 4x + 5$ slope: ____ *y*-intercept: ____

2. $x + y = 6$ slope: ____ *y*-intercept: ____

3. $y + 3 = -7x$ slope: ____ *y*-intercept: ____

4. $-x - y = -2$ slope: ____ *y*-intercept: ____

Graph Equations
pp. 434–435

Complete the organizer by following the steps given to graph an equation.

Graph: $y = -3x - 4$

Step 1: Find the slope and *y*-intercept. → **Step 1:** slope: ____ *y*-intercept: ____

Step 2: Graph the *y*-intercept at (0, –4). →

Step 3: Write the slope as $\frac{-3}{1}$. Use it to locate another point on the line. →

Step 4: Draw a line through the 2 points and extend the line. →

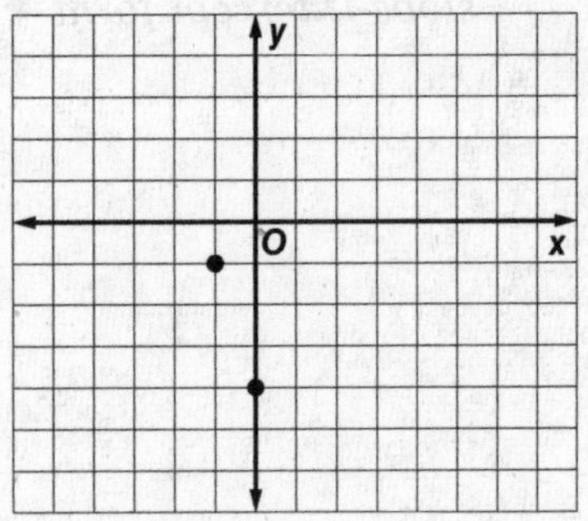

NAME ______________________ DATE __________ PERIOD __________

8-8 Writing Linear Equations

What You'll Learn

Skim Lesson 8-8. Predict two things that you expect to learn based on the headings and the Key Concept Box.

1. ______________________________________

____

2. ______________________________________

____

Active Vocabulary

Review Vocabulary Fill in each blank with the correct term or phrase. (*Lessons 8-6 and 8-7*)

slope ▶ the ratio of the ____, or ______ change, to the ____, or ________ change of a line

slope-intercept form ▶ a linear ________ in the form __________, where __ is the slope and *b* is the _________

New Vocabulary Write the definition next to the term.

point-slope form ▶ ______________________________________

____

____

Lesson 8-8

Main Idea | **Details**

Write Equations in Slope-Intercept Form
pp. 441–442, 444

Fill in each blank to write a linear equation given two points.

Given: (4, –5), and (–1, –3)

Find the slope: $m = \dfrac{\text{change of } y}{\text{change of } x} =$

Use $y - y_1 = m(x - x_1)$ form: $y - (\_\_) = \_\_\_\_ (x - (\_\_))$.

Simplify to $y - mx + b$ form: $y = \_\_\_\_ - \_\_\_\_$.

Solve Problems
p. 443

Complete the chart by summarizing the procedure.

Writing Linear Equations	
Forms	**Procedure**
from slope and *y*-intercept	
from a graph	
from two points	
from a table	

NAME ______________________ DATE __________ PERIOD __________

8-9 Prediction Equations

What You'll Learn

Scan the text in Lesson 8-9. Write two facts you learned about prediction equations as you scanned the text.

1. ______________________________

____

2. ______________________________

____

Active Vocabulary

Review Vocabulary Write the definition next to the term. *(Lesson 1-6)*

scatter plot ▶ ______________________________

____

New Vocabulary Fill in each blank with the correct term or phrase.

line of fit ▶ a _____ that is drawn on a _______ that closely approximates the ___________

Vocabulary Link In this lesson you will make *predictions* using a line or equation. *Prediction* is a word that is used in everyday English. Find the definition of *prediction* using a dictionary. Give an example of how *predictions* are used in everyday life.

____

____

____

Lesson 8-9

Main Idea	Details
Lines of Fit p. 448	**Complete the organizer about the *line of fit*.**

Line of Fit

What is the purpose?

____
____
____

What is it?

____
____
____

How do you make a line of fit?

____
____
____

How do you make predictions?

____

Fill in each blank to complete the graphic organizer for finding the equation of a line of best fit.

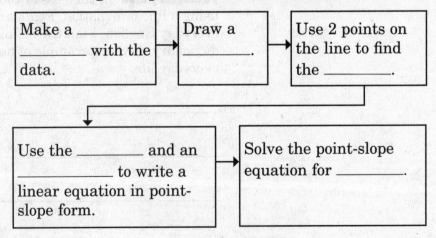

Make a ________ ________ with the data. → Draw a ________. → Use 2 points on the line to find the ________.

→ Use the ________ and an ________ to write a linear equation in point-slope form. → Solve the point-slope equation for ________.

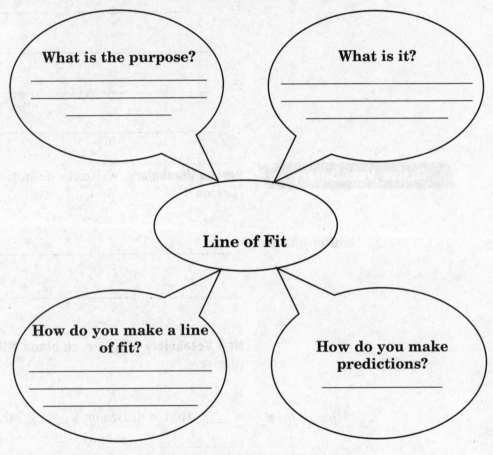

NAME ______________________ DATE __________ PERIOD __________

8-10 Systems of Equation

What You'll Learn

Scan Lesson 8-10. List two headings you would use to make an outline of this lesson.

1. ______________________________

____

2. ______________________________

____

Active Vocabulary

New Vocabulary Fill in each blank with the correct term or phrase.

system of equations ▶ a set of two or more __________ with the same __________

solving by substitution ▶ an __________ method of finding an exact __________ of a system of equations

Vocabulary Link *Substitution* is a word used in everyday English. Find the definition of *substitution* using a dictionary. Explain how the English definitions can help you remember how *substitution* is used in mathematics.

____

____

____

____

Lesson 8-10

Main Idea	Details

Solve Systems by Graphing
pp. 453–454

Compare solutions by completing the chart.

Solutions for Systems of Equations		
Solution	**Description**	**Graph**
infinite solutions	______________	
one solution	______________	
no solutions	______________	

Solve Systems by Substitution
p. 455

Fill in the diagram to complete the steps to solve a system of equations by substitution. Use the terms *variable, value, equation,* and *substitute* as often as needed.

Step 1		Step 2		Step 3
Choose one __________ and solve for one __________.	→	__________ the expression from Step 1 into the other __________. Solve for the variable.	→	Substitute the ______ for the variable found in Step 2 back into the first __________. Solve for the other variable.

NAME ______________________ DATE ____________ PERIOD ____________

CHAPTER 8

Linear Functions and Graphing

Tie It Together

Complete the graphic organizer with definitions and concepts about each topic.

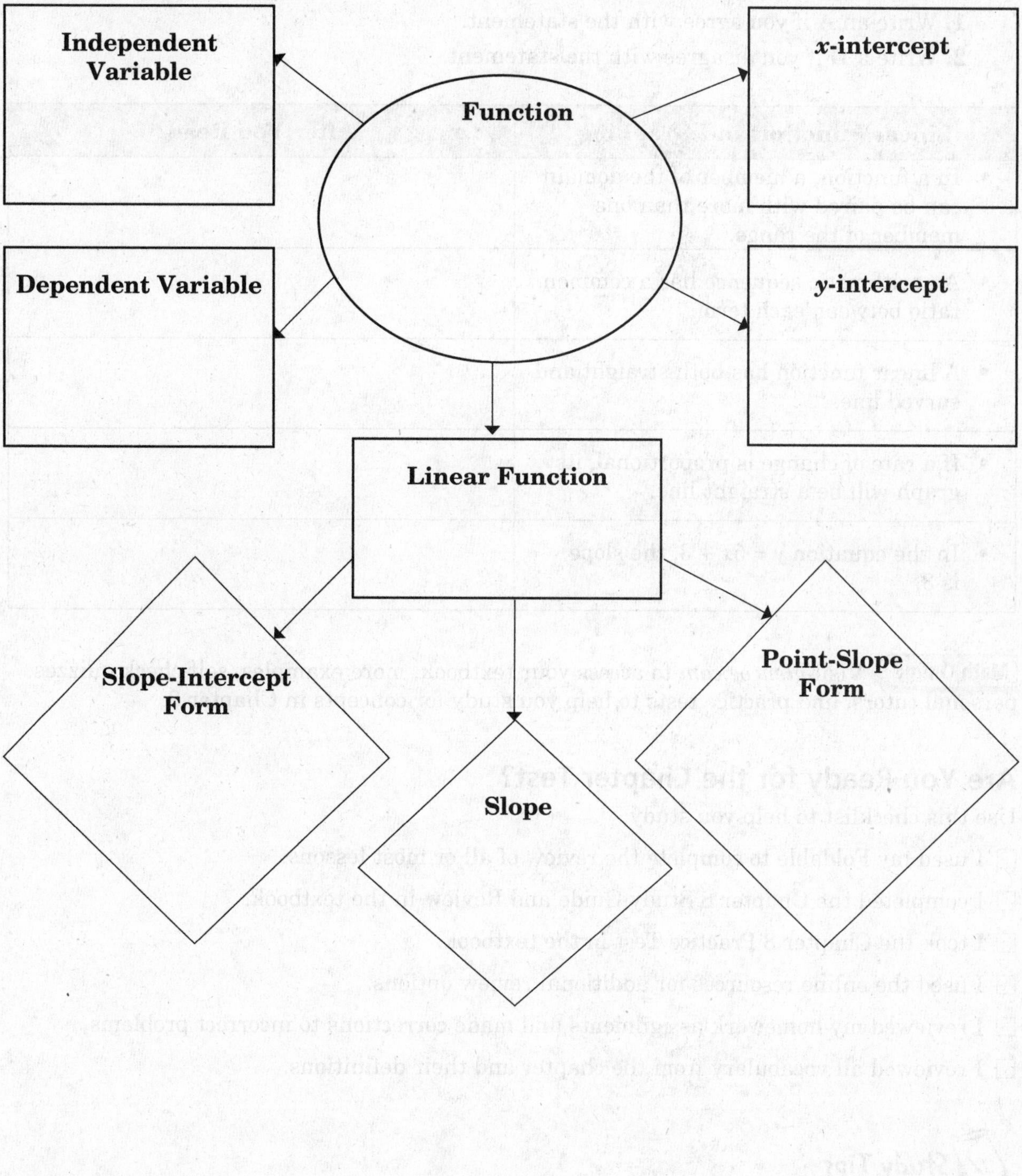

NAME ______________________________ DATE ____________ PERIOD ____________

CHAPTER 8 Linear Functions and Graphing

Before the Test

Now that you have read and worked through the chapter, think about what you have learned and complete the table below. Compare your previous answers with these.

1. Write an **A** if you agree with the statement.
2. Write a **D** if you disagree with the statement.

Linear Function and Graphing	After You Read
• In a function, a member of the domain can be paired with more than one member of the range.	
• An arithmetic sequence has a common ratio between each term.	
• A linear function has both straight and curved lines.	
• If a rate of change is proportional, its graph will be a straight line.	
• In the equation $y = 5x + 3$, the slope is 3.	

Math Online Visit *glencoe.com* to access your textbook, more examples, self-check quizzes, personal tutors, and practice tests to help you study for concepts in Chapter 8.

Are You Ready for the Chapter Test?

Use this checklist to help you study.

- ☐ I used my Foldable to complete the review of all or most lessons.
- ☐ I completed the Chapter 8 Study Guide and Review in the textbook.
- ☐ I took the Chapter 8 Practice Test in the textbook.
- ☐ I used the online resources for additional review options.
- ☐ I reviewed my homework assignments and made corrections to incorrect problems.
- ☐ I reviewed all vocabulary from the chapter and their definitions.

Study Tips

- Be an active listener in class. Take notes, circle or highlight information that your teacher stresses, and ask questions when ideas are unclear to you.

NAME ______________________ DATE __________ PERIOD __________

CHAPTER 9 Powers and Nonlinear Functions

Before You Read

Before you read the chapter, think about what you know about powers and nonlinear functions. List three things you already know about them in the first column. Then list three things you would like to learn about them in the second column.

K What I know...	W What I want to find out...

FOLDABLES® Study Organizer Construct the Foldable as directed at the beginning of this chapter.

Note Taking Tips

- **When you take notes, be sure to listen actively.**
  Always think before you write, but don't get behind in your note-taking. Remember to enter your notes legibly.
- **When you take notes, circle, underline, or star anything the teacher emphasizes.**
  When your teacher emphasizes a concept, it will usually appear on a test, so make an effort to include it in your notes.

NAME ______________________ DATE ____________ PERIOD ____________

CHAPTER 9 Powers and Nonlinear Functions

Key Points

Scan the pages in the chapter and write at least one specific fact concerning each lesson. For example, in the lesson on prime factorization, one fact might be that a monomial is a number, a variable, or a product of numbers and/or variables. After completing the chapter, you can use this table to review for your chapter test.

Lesson	Fact
9-1 Powers and Exponents	
9-2 Prime Factorization	
9-3 Multiplying and Dividing Monomials	
9-4 Negative Exponents	
9-5 Scientific Notation	
9-6 Powers of Monomials	
9-7 Linear and Nonlinear Functions	
9-8 Quadratic Functions	
9-9 Cubic and Exponential Functions	

9-1 Powers and Exponents

What You'll Learn

Skim Lesson 9-1. Predict two things that you expect to learn based on the headings and the Key Concept box.

1. ______________________________

____

2. ______________________________

____

Active Vocabulary

Review Vocabulary Write the definition next to the term. *(Lesson 1-1)*

order of operations ▶ ______________________________

____

New Vocabulary Label the diagram with the correct term.

exponent ▶

power ▶

base ▶

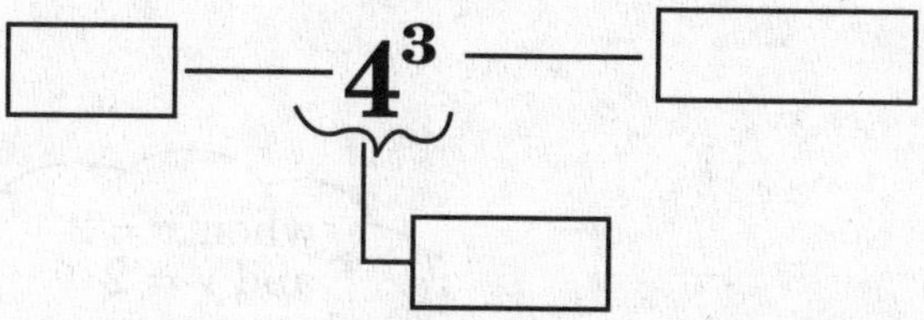

Vocabulary Link *Base* is a word that is used in everyday English. Find the definition of *base* using a dictionary. Explain how the English definition can help you remember how *base* is used in mathematics.

____

____

____

Main Idea	Details

Use Exponents
pp. 471–472

Fill in the blank for each verbal expression with a numeric expression with exponents.

1. 8 to the seventh power ______

2. 3 cubed ______

3. 6 to the fourth power ______

4. 4 to the first power ______

5. 7 squared ______

Evaluate Expressions
pp. 472–473

Complete the organizer to evaluate the expression with the values given for *x* and *y*.

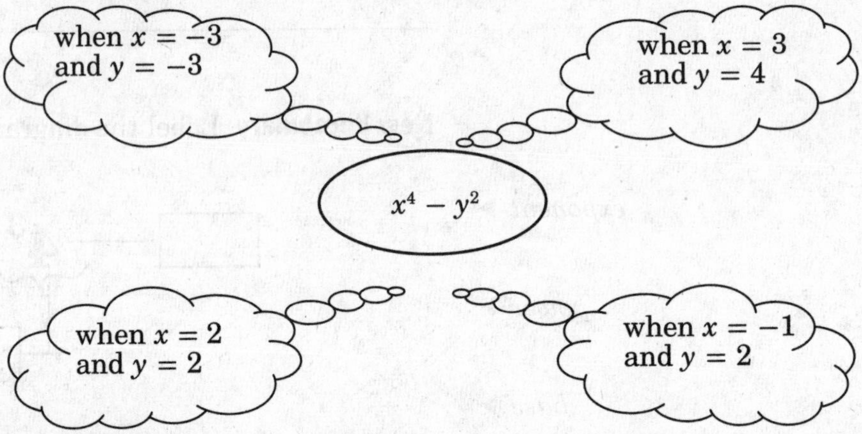

Helping You Remember

A classmate states that $3^2 = 6$. How would you explain the correct solution? Use words, drawings, or models in your explanation.

____

____

____

NAME ______________________ DATE ____________ PERIOD ____________

9-2 Prime Factorization

What You'll Learn

Scan Lesson 9-2. List two headings you would use to make an outline of this lesson.

1. ______________________

____

2. ______________________

____

Active Vocabulary

New Vocabulary Match the term with its definition by drawing a line to connect the two.

monomial	when a composite number is expressed as the product of prime factors
composite number	to write a number as a product of its factors
factor tree	a whole number with exactly two unique factors, 1 and itself
factor	an expression that is a number, a variable, or a product of numbers and variables
prime number	a way to find the prime factorization of a number
prime factorization	a whole number that has more than two factors

Vocabulary Link *Composite* is a word that is used in everyday English. Find the definition of *composite* using a dictionary. Explain how the English definition can help you remember how a *composite number* is used in mathematics.

____

____

____

Lesson 9-2

NAME ______________________ DATE __________ PERIOD __________

Main Idea	**Details**

Write Prime Factorization
pp. 476–477

Complete the factor tree.

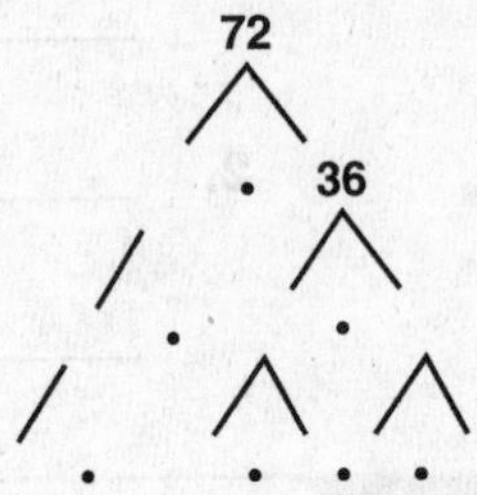

The prime factorization of 72 is ________.

Factor Monomials
pp. 477–478

Fill in each blank with the monomial whose factors are shown.

1. 2 3 3 x y y ________

2. –1 5 5 a a a ________

3. 3 7 11 s s s s s ________

4. –1 x x x ________

Helping You Remember

Explain the relationship between the terms *base*, *exponent*, and *power*.

____

____

____

NAME ______________________ DATE __________ PERIOD __________

9-3 Multiplying and Dividing Monomials

What You'll Learn

Scan the text under the *Now* heading. List two things you will learn about in the lesson.

1. ____________________

____

2. ____________________

____

Active Vocabulary

Review Vocabulary Write the term next to the definition. (*Lessons 1-3 and 2-1*)

____________ ▶ a number greater than zero

____________ ▶ a number less than zero

____________ ▶ the whole numbers and their opposites

____________ ▶ The order numbers are multiplied does not change the product.
____

____________ ▶ The order numbers are grouped does not change the sum.
____

Finish each property.

Product of Powers Property

$a \quad \cdot a^n = a^m$

Quotient of Powers Property

$a^m \div a \quad = a \quad\quad n$

Lesson 9-3

Main Idea	Details
Multiply Monomials pp. 481–482	**Fill in the blanks to find each product.**

1. $4^3 \quad 4^2 = 4^{\square+\square} = 4^{\square}$
2. $2^5 \quad 2^3 = 2^{\square+\square} = \square^{\square}$
3. $5^3 \quad 5^4 = 5^{\square+\square} = \square^{\square}$
4. $2y^3 \quad -3y^3 = \square\,\square^{\square+\square} = \square\,\square^{\square}$
5. $5x^4 \quad 3x^3 = \square\,\square^{\square+\square} = \square\,\square^{\square}$

Divide Monomials pp. 482–483

Cross out the one that does not belong. Then state the relationship among the three remaining parts of the circle.

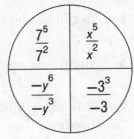

$\frac{7^5}{7^2}$	$\frac{x^5}{x^2}$
$\frac{-y^6}{-y^3}$	$\frac{-3^3}{-3}$

The relationship is:

____.

Helping You Remember

Restate the Product of Powers Property and the Quotient of Powers Property in your own words.

____

____

____

____

____

NAME ______________________ DATE __________ PERIOD __________

9-4 Negative Exponents

What You'll Learn

Skim Lesson 9-4. Predict two things you expect to learn based on the headings and the Key Concept box.

1. ______________________________

____

2. ______________________________

____

Active Vocabulary

Review Vocabulary Write the definitions next to each term. *(Lessons 1-1, 1-2, 1-3 and 9-1)*

deductive reasoning ▶ ______________________________

____

exponent ▶ ______________________________

____

power ▶ ______________________________

____

base ▶ ______________________________

____

evaluate ▶ ______________________________

____

algebraic expression ▶ ______________________________

____

Main Idea	Details
Negative Exponents pp. 486–487	**Fill in each blank to prove $y^{-3} = \frac{1}{y^3}$.** Start with $\frac{y^4}{y^7}$.

$\frac{y^4}{y^7} \longrightarrow$ Using the ______________________, the quotient is $\square^{\square - \square} = \square^{\square}$.

$\frac{y^4}{y^7} \longrightarrow$ By ______________, the quotient is $\frac{y \cdot y \cdot y \cdot y}{y \cdot y \cdot y \cdot y \cdot y \cdot y \cdot y} = \frac{\square}{\square}$.

Because $\frac{y^4}{y^7}$ is equal to both $\square^{\square}$ and $\frac{\square}{\square}$, $\square = \square$.

Evaluate Expressions p. 488

Fill in the diagram to complete the steps to evaluate an expression with negative exponents. Use the terms *order of operations, positive, replace,* and *simplify*.

Step 1	Step 2	Step 3
______________ the value given for the variable. →	Write the negative power as a ________ power. →	Use the ________ ____________ and ____________.

Helping You Remember

Explain how negative exponents can be written as positive exponents.

____

____

NAME ______________________________ DATE ____________ PERIOD ____________

9-5 Scientific Notation

What You'll Learn

Scan the text in Lesson 9-5. Write two facts you learned about scientific notation as you scanned the text.

1. ____

____

2. ____

____

Active Vocabulary

New Vocabulary Match the following terms with the correct examples by drawing a line to connect the two.

	0.000050
standard form	
	5.0×10^5
	2.8×10^3
scientific notation	
	3,700
	8,900,000,000

Vocabulary Link *Standard* is a word that is used in everyday English. Find the definition of *standard* using a dictionary. Explain how the English definition can help you remember how *standard form* is used in mathematics.

____

____

____

____

____

Lesson 9-5

Main Idea | Details

Scientific Notation
pp. 493–494

Complete the organizer about *scientific notation.*

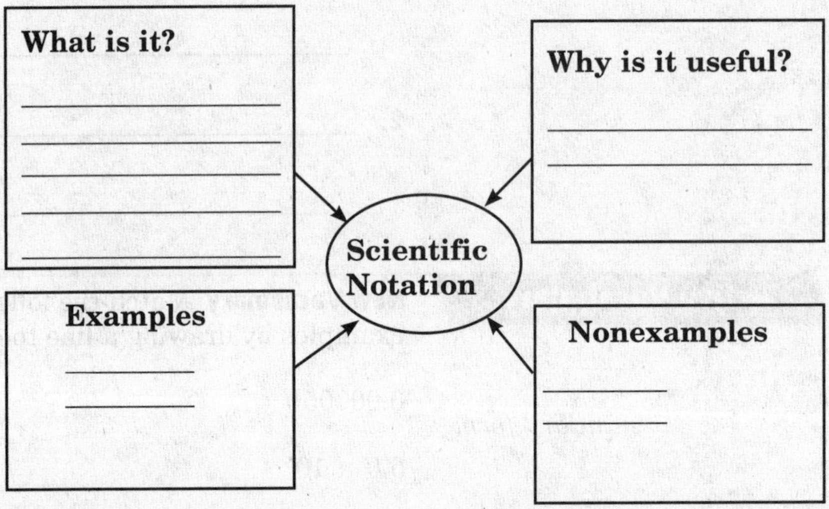

Compare and Order Numbers
p. 495

Write the numbers in order from greatest to least.

1. 4.05×10^5, 4.2×10^5, 3.0×10^5 , 1.3×10^5

2. 2.4×10^{-3}, 2.0×10^{-2}, 3.1×10^3, 2.9×10^{-2}

Helping You Remember

Explain how to express a number greater than 1, a number less than 1, and then the number 1 in scientific notation.

___

___

___

___

___

NAME ______________________________ DATE __________ PERIOD __________

9-6 Powers of Monomials

What You'll Learn

Skim the lesson. Write two things you already know about powers of monomials.

1. ______________________________

____

2. ______________________________

____

Active Vocabulary

Review Vocabulary Fill in each blank with the correct term or phrase. *(Lessons 9-1 and 9-2)*

monomial ▶ an expression that is a number, a __________, or a __________ of numbers and or variables

power ▶ a __________ that is expressed using an __________

Vocabulary Link Write a *power* that has a *base* of 7 and an *exponent* of 4. Then write x to the fifth *power*. Write y *squared*. Write a monomial that is the product of the number 2 and k cubed.

____

____

Finish for each property.

Power of a __________ Property

$$(a^m)^n = a^{m\_n}$$

Power of a __________ Property

$$(ab)^m = a^{\_}b^{\_}$$

Lesson 9-6

Main Idea	Details

Power of a Power
p. 499

Fill in the blanks with each product.

1. $(5^3)^2 = 5^{\square \cdot \square} = 5^{\square}$

2. $(x^5)^4 = x^{\square \cdot \square} = x^{\square}$

3. $(6^2)^{-2} = 6^{\square \cdot \square} = 6^{\square}$ or $\frac{\square}{\square}$

4. $(y^{-3})^{-4} = y^{\square \cdot \square} = y^{\square}$

Power of a Product
pp. 500–501

Compare the two properties of powers by filling out the chart.

	Power of Powers	Power of a Product
Why?		
How?		
Example		

Helping You Remember

Compare and contrast the Quotient of Powers Property and the Product of Powers Property.

____

____

____

____

____

NAME ______________________ DATE __________ PERIOD __________

9-7 Linear and Nonlinear Function

What You'll Learn

Skim the Examples for Lesson 9-7. Predict two things you think you will learn about linear and nonlinear function.

1. ______________________________

____

2. ______________________________

____

Active Vocabulary

Review Vocabulary Write the definition next to each term. *(Lesson 1-5)*

function ▶ ______________________________

____

function rule ▶ ______________________________

____

function table ▶ ______________________________

____

New Vocabulary Fill in each blank with the correct term or phrase.

nonlinear functions ▶ functions that ________ have constant ______________, therefore their graphs are *not* ______________

Lesson 9-7

Main Idea | **Details**

Graphs of Nonlinear Functions
p. 504

Model a *linear* and *nonlinear* function on the coordinate planes. On the lines below, write a real-life example of each kind of function.

Linear Function

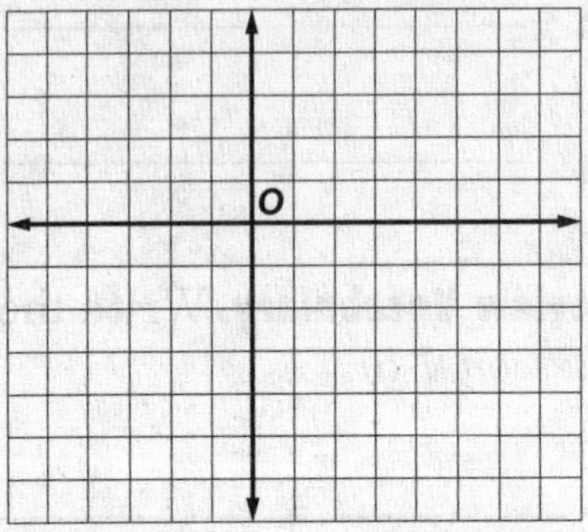

Nonlinear Function

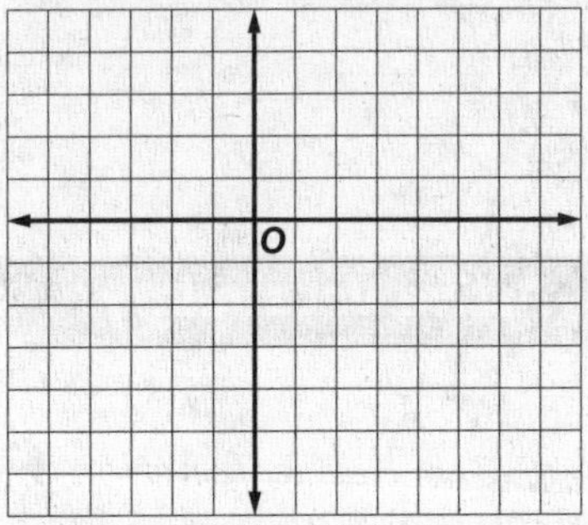

____

____

____

____

____

Equations and Tables
pp. 505–506

Complete the organizer to summarize three ways to determine if a function is *linear or nonlinear*.

Is the graph a straight line?
- **Yes** → ____________
- **No** → ____________

Can the equation of the line be written in the form $y = mx + b$?
- **Yes** → ____________
- **No** → ____________

In a function table, are the changes in x and y constant?
- **Yes** → ____________
- **No** → ____________

9-8 Quadratic Functions

What You'll Learn

Scan Lesson 9-8. List two headings you would use to make an outline of this lesson.

1. ______________________

____

2. ______________________

____

Active Vocabulary

New Vocabulary Write the definition next to each term.

parabola ▶ ______________________

____

quadratic function ▶ ______________________

____

Vocabulary Link A *parabola* is the shape that is seen in everyday life. Give an example of something that has a parabola shape in real life.

____

____

____

Main Idea | **Details**

Graph Quadratic Function

pp. 510–511

Complete the organizer by filling in the blanks. Then complete the example.

Graph a Quadratic Function

Step 1:
Make a ______________.

Step 2:
_____the ordered pairs.

Step 3:
________ the points with a curved ____.

Graph: $y = 2x^2 + 3$

Step 1:

x	$y = 2x^2 + 3$	y

Step 2 and Step 3:

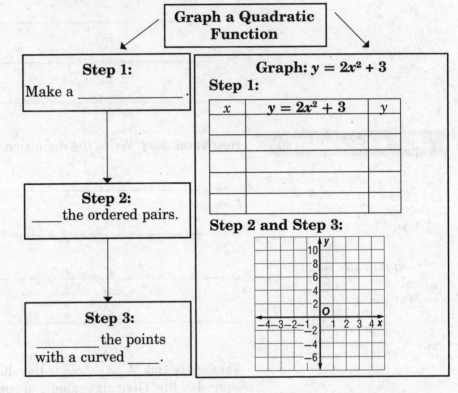

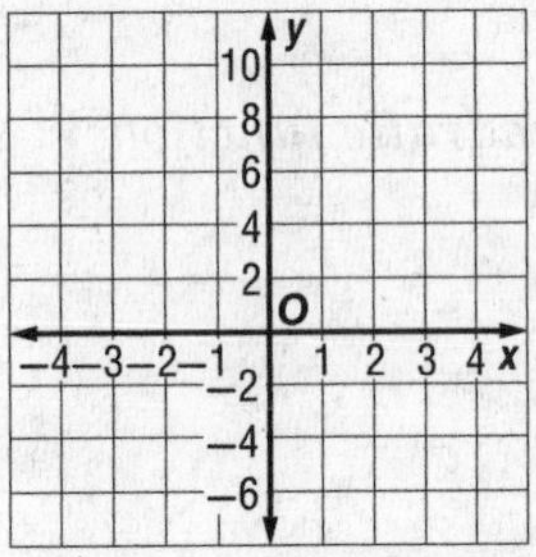

Use Quadratic Function

p. 511

Fill in the blanks by using the information below.

A ball is thrown into the sky. The equation that gives the ball's height in meters h as a function of time t is $h = -4.9t^2 + 12t + 3$.

1. What is the height of the ball after $t = 1$ second?

____

2. How high is the ball after 2 seconds? __________

3. What can you say about the ball's path between 1 and 2 seconds? ____

________________.

9-9 Cubic and Exponential Functions

What You'll Learn

Scan the text under the *Now* heading. List two things you will learn about this lesson.

1. ______________________________

____

2. ______________________________

____

Active Vocabulary

Review Vocabulary Fill in each blank with the correct term or phrase. (*Lesson 9-8*)

quadratic function ▶ a function that can be written in the form ____________, where $a \neq 0$

parabola ▶ the graph of a ____________ function

New Vocabulary Match the term with the correct form by drawing a line to connect the two.

cubic function $y = a^x + c$, where $a \neq 0$, $a \neq 1$

exponential function $y = ax^3 + bx^2 + cx + d$, where $a \neq 0$

Vocabulary Link *Exponential* is a word that is used in everyday English. Find the definition of *exponential* using a dictionary. Explain how the English definition can help you remember the shape of the graph of an *exponential function* in mathematics.

____

____

____

Main Idea	Details

Cubic Functions
pp. 516–517

Fill in the organizer for *cubic functions*.
are given.

What is a cubic function?	Sketch the shape of a graph of a cubic function.
Examples of Cubic Functions	**Nonexamples of Cubic Functions**

Cubic Functions

Exponential Functions
pp. 517–518

Fill in each blank with the value of *y*.

1. $y = 2^x - 1$, when $x = 3$: (3, ____)

2. $y = 4^x + 2$, when $x = -1$: (−1, ____)

3. $y = 2^x - 3$, when $x = 2$: (2, ____)

4. $y = 5^x$, when $x = -2$: (−2, ____)

Helping You Remember

You have learned to graph quadratic and cubic functions. Make a list of the steps you use to graph the two functions.

____

____

NAME ________________ DATE ________ PERIOD ________

CHAPTER 9 Powers and Nonlinear Functions

Tie It Together

Complete the graphic organizer by writing an equivalent form of the exponential expression.

__________ $\rightarrow x^a \leftarrow$ __________

$x^m \cdot x^n$	
$\frac{x^m}{x^n}$	
$(x^m)^n$	
$(xy)^m$	
x^{-m}	
x^0, when $x \neq 0$	

Complete the graphic organizer with types of functions and their general equations.

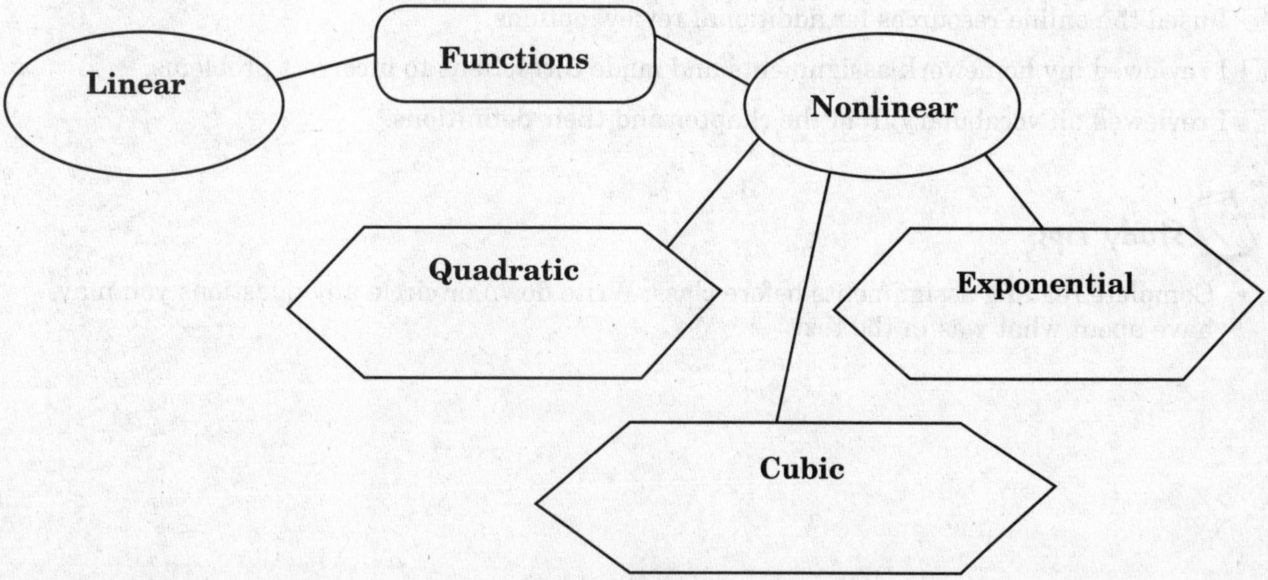

NAME ______________________ DATE __________ PERIOD __________

CHAPTER 9

Powers and Nonlinear Functions

Before the Test

Review the ideas you listed in the table at the beginning of the chapter. Cross out any incorrect information in the first column. Then complete the table by filling in the third column.

K What I know...	W What I want to find out...	L What I learned...

Math Online Visit *glencoe.com* to access your textbook, more examples, self-check quizzes, personal tutors, and practice tests to help you study for concepts in Chapter 9.

Are You Ready for the Chapter Test?

Use this checklist to help you study.

☐ I used my Foldable to complete the review of all or most lessons.

☐ I completed the Chapter 9 Study Guide and Review in the textbook.

☐ I took the Chapter 9 Practice Test in the textbook.

☐ I used the online resources for additional review options.

☐ I reviewed my homework assignments and made corrections to incorrect problems.

☐ I reviewed all vocabulary from the chapter and their definitions.

Study Tips

- Complete reading assignments before class. Write down or circle any questions you may have about what was in the text.

NAME ______________________ DATE __________ PERIOD __________

CHAPTER 10

Real Numbers and Right Triangles

Before You Read

Before you read the chapter, think about what you know about real numbers and right triangles. List three things you already know about them in the first column. Then list three things you would like to learn about them in the second column.

K What I know...	W What I want to find out...

FOLDABLES® Study Organizer Construct the Foldable as directed at the beginning of this chapter.

Note Taking Tips

- **Before going to class, look over your notes from the previous class, especially if the day's topic builds from the last one.**
- **When you take notes, write down the math problem and each step in the solution using math symbols.**

  Next to each step, write down, in your own words, exactly what you are doing.

NAME ______________________ DATE __________ PERIOD __________

CHAPTER 10

Real Numbers and Right Triangles

Key Points

Scan the pages in the chapter and write at least one specific fact concerning each lesson. For example, in the lesson on triangles, one fact might be that a vertex is a point where line segments intersect. After completing the chapter, you can use this table to review for your chapter test.

Lesson	Fact
10-1 Squares and Square Roots	
10-2 The Real Number System	
10-3 Triangles	
10-4 The Pythagorean Theorem	
10-5 The Distance Formula	
10-6 Special Right Triangles	

10-1 Squares and Square Roots

What You'll Learn

Scan the text in Lesson 10-1. Write two facts you learned about squares and square roots as you scanned the text.

1. ____

____

2. ____

____

Active Vocabulary

New Vocabulary Match the term with its definition by drawing a line to connect the two.

perfect square	one of a number's two equal factors
square root	indicates a positive square root
radical sign	a rational number whose square root is a whole number

Main Idea / Details

Find Square Roots
p. 537

Cross out the square root in the concept circle that does not belong. Then describe the relationship of the remaining three parts.

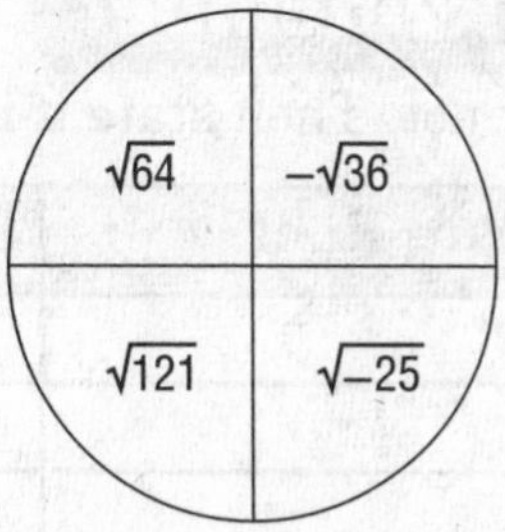

The relationship is ____

____.

Lesson 10-1 *(continued)*

Main Idea	Details

Estimate Square Roots
pp. 538–539

Complete the organizer by following the steps to estimate a square root. Then complete the example.

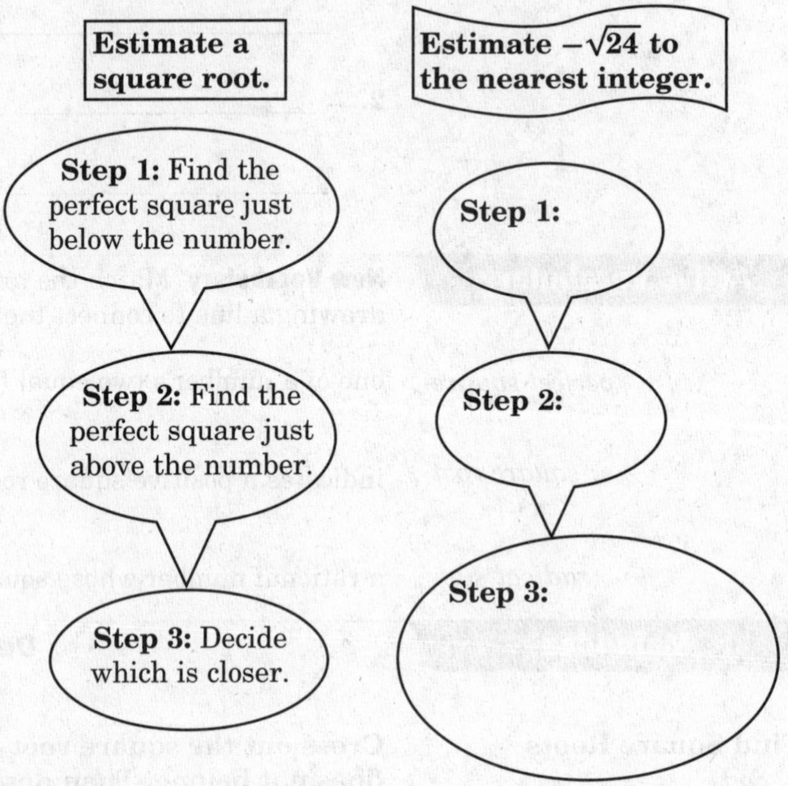

Estimate a square root.
Step 1: Find the perfect square just below the number.
Step 2: Find the perfect square just above the number.
Step 3: Decide which is closer.
Estimate $-\sqrt{24}$ to the nearest integer.
Step 1:
Step 2:
Step 3:

Helping You Remember

Tell whether each number has a square root and explain why or why not. Then state if it is a perfect square and explain.

	Real Square Root?	Perfect Square?
26		
–81		
256		
2500		
–5		

NAME ______________________ DATE ____________ PERIOD ____________

10-2 The Real Number System

What You'll Learn

Scan the text under the *Now* heading. List two things you will learn about the real number system.

1. ____

2. ____

Active Vocabulary

New Vocabulary Label the diagram with the correct term.

irrational numbers ▶

real numbers ▶

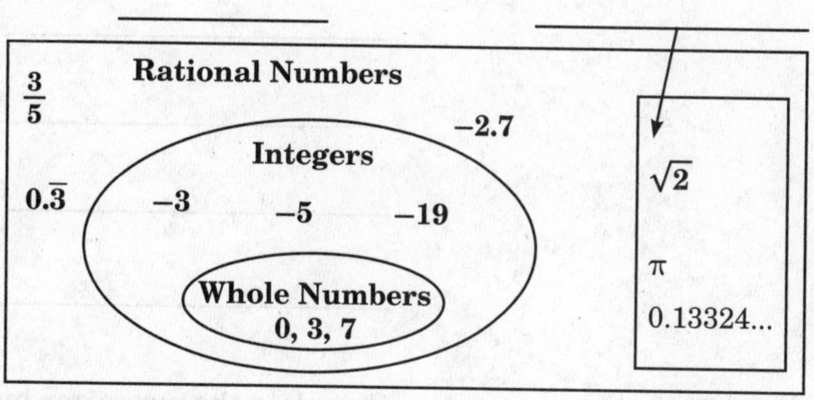

Vocabulary Link *Irrational* is a word that is used in everyday English. Find the definition of *irrational* using a dictionary. Explain how the English definition can help you remember how *irrational* is used in mathematics.

____

____

____

Lesson 10-2

Lesson 10-2 *(continued)*

Main Idea	Details

Identify and Compare Real Numbers

pp. 543–544

Model and explain how to use a number line to list $\frac{7}{8}$, $0.\overline{67}$, $\sqrt{3}$, and $\frac{3}{4}$ from least to greatest. Write an inequality and explanation on the lines below.

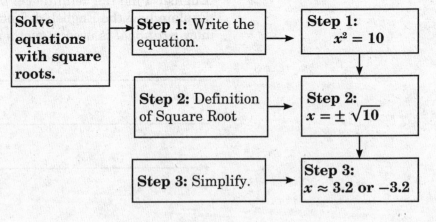

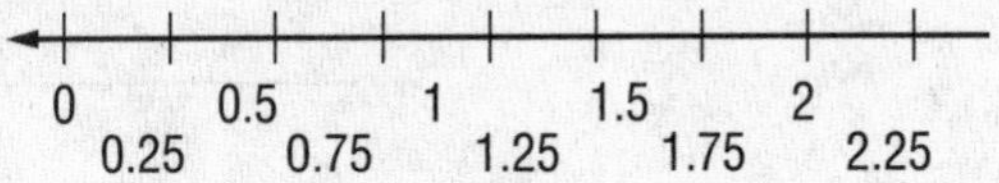

____

____

____

____

Solve Equations

p. 545

Complete the organizer by following the steps to solve the equation $x^2 = 10$ using the definition of a square root.

Solve equations with square roots. → **Step 1:** Write the equation. → **Step 1:** $x^2 = 10$

Step 2: Definition of Square Root → **Step 2:** $x = \pm\sqrt{10}$

Step 3: Simplify. → **Step 3:** $x \approx 3.2$ or -3.2

NAME ______________________ DATE ____________ PERIOD ____________

10-3 Triangles

What You'll Learn

Skim Lesson 10-3. Predict two things that you expect to learn based on the headings and the Key Concept boxes.

1. ______________________________

____

2. ______________________________

____

Active Vocabulary

New Vocabulary Write the definition next to each term.

congruent ▶ ______________________________

triangle ▶ ______________________________

____

vertex ▶ ______________________________

____

line segment ▶ ______________________________

____

Main Idea

Details

Find Angle Measures
pp. 550–551

Fill in the diagram to determine the angle measures of $\triangle KLM$ with a ratio of 1:3:1.

Step 1		Step 2		Step 3
Use x to represent the measure of the ______ and ______ ______.	→	Use ______ to represent the measure of the ______.	→	Write the equation, ______ ______. Solve for x.

Because $x =$ ______, the measure of the angles are ______ ______.

Lesson 10-3

Main Idea	Details

Classify Triangles
pp. 551–552

Summarize information about triangles in the graphic organizer. Sample answers are given.

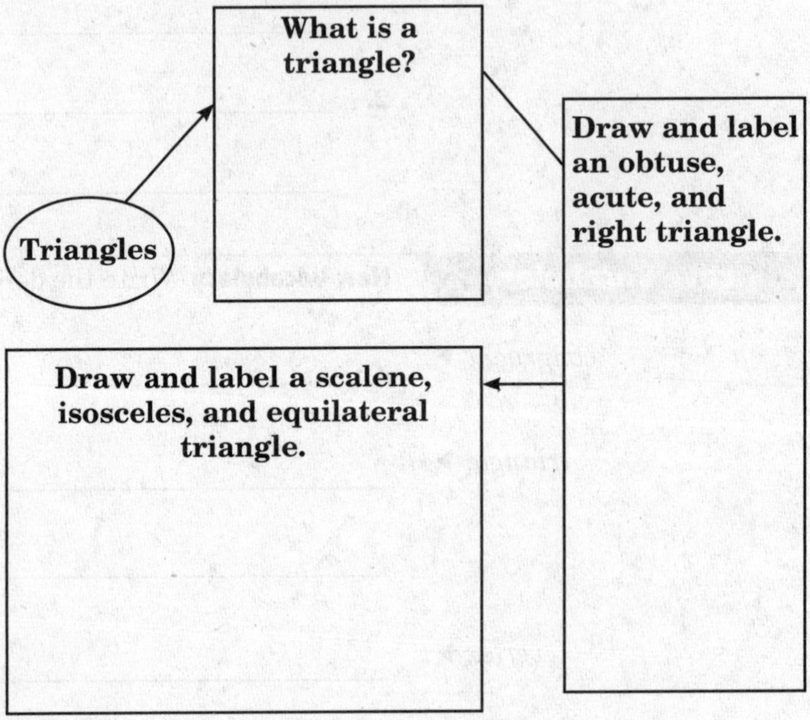

Helping You Remember

Describe an obtuse, scalene triangle.

____

____

Describe an equilateral triangle. ______________________________

____

NAME ______________________ DATE ____________ PERIOD ____________

10-4 The Pythagorean Theorem

What You'll Learn

Skim the Examples in Lesson 10-4. Predict two things you think you will learn about the Pythagorean Theorem.

1. ______________________

____

2. ______________________

____

____

Active Vocabulary

New Vocabulary Write the definition next to each term.

hypotenuse ▶ ______________________

____

Pythagorean Theorem ▶ ______________________

____

____

legs ▶ ______________________

solving a right triangle ▶ ______________________

____

____

converse of the Pythagorean Theorem ▶ ______________________

____

Lesson 10-4

Main Idea	Details

Use the Pythagorean Theorem
pp. 558–559

Fill in the organizer about using the *Pythagorean Theorem.*

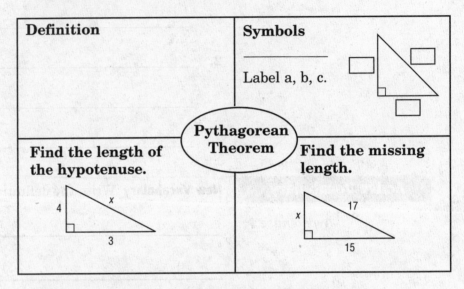

Pythagorean Theorem

Definition

Symbols

____

Label a, b, c.

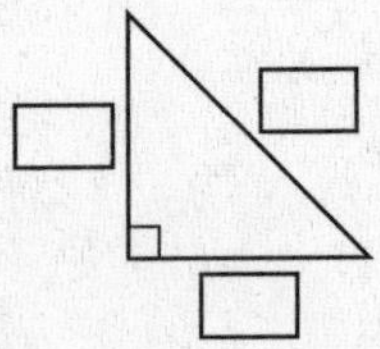

Find the length of the hypotenuse.

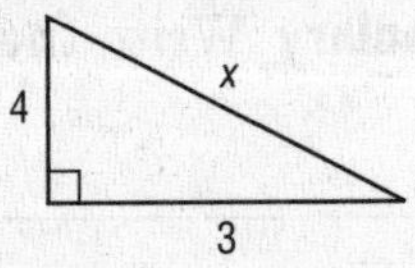

Find the missing length.

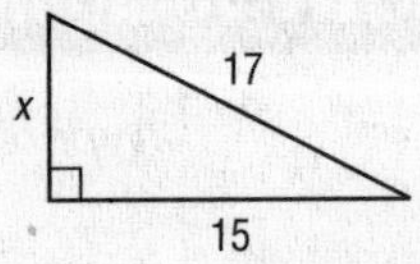

Use the Converse of the Pythagorean Theorem
p. 560

Fill in the blanks and determine whether each triangle is a right triangle.

1. 12 in., 35 in., 37 in.

$a^2 + b^2 = c^2$

$\square + \square \stackrel{?}{=} \square$

$\square + \square \stackrel{?}{=} \square$

$\square = \square$

Is it a right triangle?

2. 7 cm, 23 cm, 24 cm

$a^2 + b^2 = c^2$

$\square + \square \stackrel{?}{=} \square$

$\square + \square \stackrel{?}{=} \square$

$\square \neq \square$

Is it a right triangle?

NAME ______________________ DATE __________ PERIOD __________

10-5 The Distance Formula

What You'll Learn

Skim the lesson. Predict two things that you expect to learn based on the headings and the Key Concept boxes.

1. ______________________________

____

2. ______________________________

____

Active Vocabulary

Review Vocabulary Label the diagram with the correct term. *(Lessons 10-3 and 10-4)*

hypotenuse ▶

leg ▶

vertex ▶

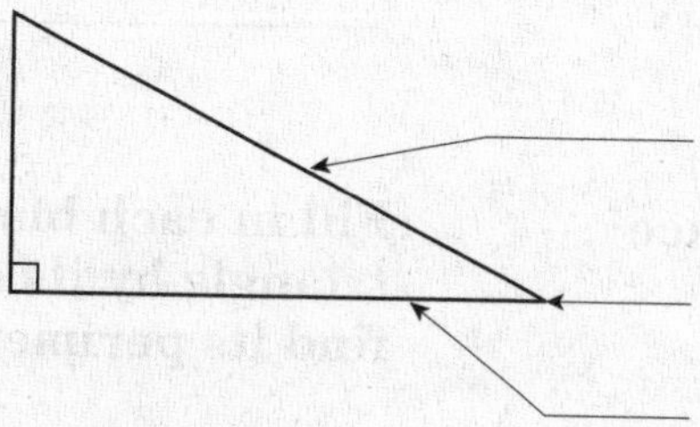

New Vocabulary Fill in each blank with the correct term or phrase.

distance formula ▶ The distance d between __________ with coordinates ________ and ________ is given by the formula

$d =$ ____________________

Lesson 10-5

Main Idea	Details

Find the Distance Between Points
pp. 565–566

Model the solution to find the distance between the points on the coordinate plane. Use the lines to show your calculations. Round to the nearest tenth if necessary.

$(x_1, y_1) = (-4, 3); (x_2, y_2) = (5, 5)$

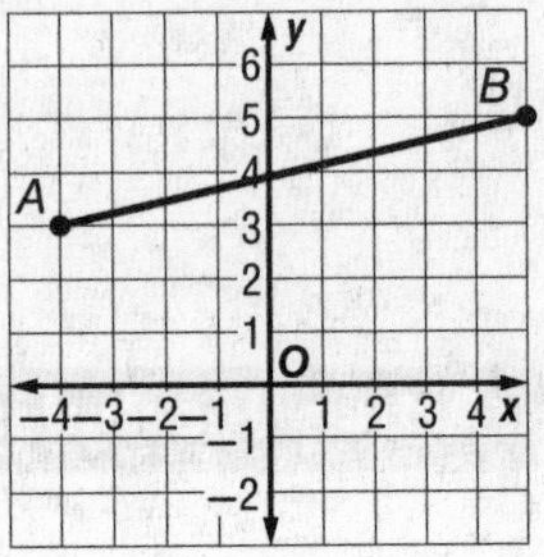

AB ______________________________

____

____

____

Apply the Distance Formula
p. 567

Fill in each blank to describe the steps to classify a triangle by its sides on a coordinate plane and then find its perimeter.

Step 1		Step 2		Step 3
Use the distance formula, ________ to find the ______ of each side.	→	________ the triangle as scalene, equilateral, or isosceles using the ________.	→	Find the ______ of the triangle's ________ to find the ________.

Helping You Remember

Describe how you would find the perimeter of $\triangle STU$. List any formulas that must be used.

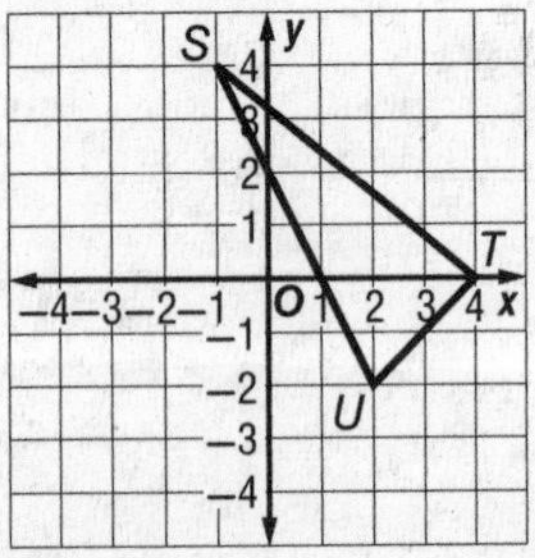

____

____

____

____

NAME ______________________ DATE ____________ PERIOD ____________

10-6 Special Right Triangles

What You'll Learn

Scan Lesson 10-6. List two headings you would use to make an outline of this lesson.

1. ______________________________________

____

2. ______________________________________

____

Active Vocabulary

Review Vocabulary Write the term next to each definition. *(Lessons 6-7, 10-1, 10-3, and 10-4)*

______________ ▶ formed by three line segments that intersect only at their endpoints

______________ ▶ the point where two line segments that form a side of a triangle meet

______________ ▶ figures that have the same shape but not necessarily the same size

Main Idea | Details

Find Measures in 45 –45 –90 Triangles
pp. 571–572

Use △*ABC* to fill in each blank.

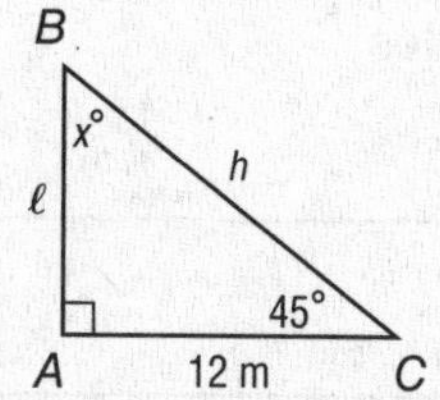

1. The measure of $\angle B$ is _____ because ______________________________________.

2. The length of the hypotenuse, *h*, is _____ because ______________________________________.

3. The length of side, *l*, is 12 meters because ______________________________________.

Main Idea	Details

Find Measures in 30 –60 –90 Triangles

pp. 572–573

Compare 45 –45 90 and 30 60 90 triangles by filling in each blank of the organizer.

Special Right Triangles

↓ **45 –45 –90 Triangles**

The length of the hypotenuse is ___ times the length of ___ leg.

The length of the two legs are _______.

↓ **30 –60 –90 Triangles**

The length of the hypotenuse is _ times the length of the _______ leg.

The length of the _______ leg is ___ times the length of the _______ leg.

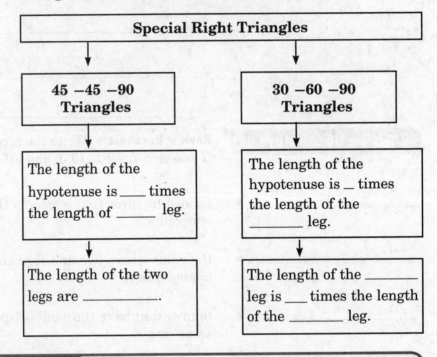

Helping You Remember

Describe the properties of a 30° 60°–90° triangle. Include the ways that students sometimes mismeasure the side lengths or angles.

____

____

____

____

____

____

NAME ______________________ DATE __________ PERIOD __________

CHAPTER 10 Real Numbers and Right Triangles

Tie It Together

Sketch an example of each type of triangle if possible. If the sketch is not possible mark an X in the box.

Angle Measure/ Side Length	Acute	Obtuse	Right
Equilateral			
Isosceles			
Scalene			

NAME ______________________ DATE ____________ PERIOD ____________

CHAPTER 10

Real Numbers and Right Triangles

Before the Test

Review the ideas you listed in the table at the beginning of the chapter. Cross out any incorrect information in the first column. Then complete the table by filling in the third column.

K What I know...	W What I want to find out...	L What I learned...

Math Online Visit *glencoe.com* to access your textbook, more examples, self-check quizzes, personal tutors, and practice tests to help you study for concepts in Chapter 10.

Are You Ready for the Chapter Test?

Use this checklist to help you study.

☐ I used my Foldable to complete the review of all or most lessons.

☐ I completed the Chapter 10 Study Guide and Review in the textbook.

☐ I took the Chapter 10 Practice Test in the textbook.

☐ I used the online resources for additional review options.

☐ I reviewed my homework assignments and made corrections to incorrect problems.

☐ I reviewed all vocabulary from the chapter and their definitions.

Study Tips

- Use abbreviations while note-taking to use less time and room. Write neatly and place a question mark by any information that you do not understand.

NAME ______________________ DATE ____________ PERIOD ____________

CHAPTER 11 Distance and Angle

Before You Read

Before you read the chapter, respond to these statements.

1. Write an **A** if you agree with the statement.
2. Write a **D** if you disagree with the statement.

Before You Read	Distance and Angle
	• Parallel lines intersect to form right angles.
	• A figure that rotates about a fixed point does not change shape or size.
	• A quadrilateral is a polygon that has four sides.
	• An example of a polygon is a circle.
	• The formula to find the circumference of a circle is $C = 2\pi r$.

FOLDABLES® Study Organizer Construct the Foldable as directed at the beginning of this chapter.

Note Taking Tips

- **It is helpful to read through your notes before beginning your homework.** Look over any page referenced material.
- **As soon as possible, go over your notes.** Clarify any ideas that were not complete.

NAME ______________________ DATE __________ PERIOD __________

CHAPTER 11 Distance and Angle

Key Points

Scan the pages in the chapter and write at least one specific fact concerning each lesson. For example, in the lesson on quadrilaterals, one fact might be that the sum of the measures of the angles of a quadrilateral is 360°. After completing the chapter, you can use this table to review for your chapter test.

Lesson	Fact
11-1 Angle and Line Relationships	
11-2 Congruent Triangles	
11-3 Rotations	
11-4 Quadrilaterals	
11-5 Polygons	
11-6 Area of Parallelograms, Triangles, and Trapezoids	
11-7 Circles and Circumference	
11-8 Area of Circles	
11-9 Area of Composite Figures	

11-1 Angle and Line Relationships

What You'll Learn

Skim Lesson 11-1. Predict two things that you expect to learn based on the headings and the Key Concept boxes.

1. ________________________________

2. ________________________________

Active Vocabulary

New Vocabulary Write the definition next to each term.

vertical angles ▶ ________________________________

adjacent angles ▶ ________________________________

complementary angles ▶ ________________________________

supplementary angles ▶ ________________________________

perpendicular lines ▶ ________________________________

parallel lines ▶ ________________________________

transversal ▶ ________________________________

alternative interior angles ▶ ________________________________

alternative exterior angles ▶ ________________________________

corresponding angles ▶ ________________________________

Main Idea	Details

Angle Relationships
pp. 589–590

Complete the model so that ∠*ABC* is complementary to ∠*ABD* and ∠*ABC* is supplementary to ∠*ABD*. Label each angle measure.

Complementary Angles	Supplementary Angles
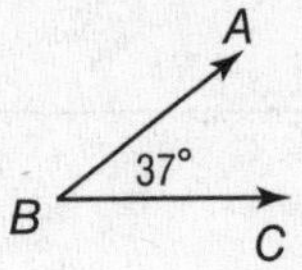	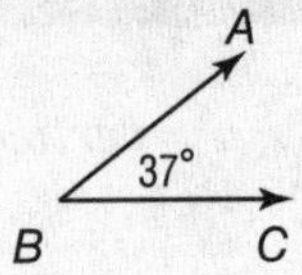

Parallel Lines
pp. 590–591

Draw a transversal *t* which intersects with two parallel lines *a* and *b*. Label all interior angles, exterior angles, alternative interior and exterior angles, and corresponding angles.

Helping You Remember

Look up the meaning of the prefix *trans-* in the dictionary. Write down four words that have *trans-* as a prefix. How can the meaning of the prefix help you remember the meaning of transversal?

____

____

NAME ______________________ DATE __________ PERIOD __________

11-2 Congruent Triangles

What You'll Learn

Scan Lesson 11-2. List two headings you would use to make an outline of this lesson.

1. ______________________________

____

2. ______________________________

____

Active Vocabulary

Review Vocabulary Write the correct term next to each definition. (*Lesson 10-3*)

______________ ▶ formed by three line segments that intersect only at their endpoints

______________ ▶ the point where two line segments that form a side of a triangle meet

______________ ▶ the part of a line containing two endpoints and all of the points between them

New Vocabulary Fill in each blank with the correct term or phrase.

congruent ▶ ______________ that have the same ______________ and ______________ are congruent.

corresponding parts ▶ The ______________ of ______________ triangles that ______________ or correspond are called corresponding parts.

Lesson 11-2

Main Idea	Details

Corresponding Parts
pp. 598–600

Fill in each blank to complete the congruence statements for the congruent triangles below.

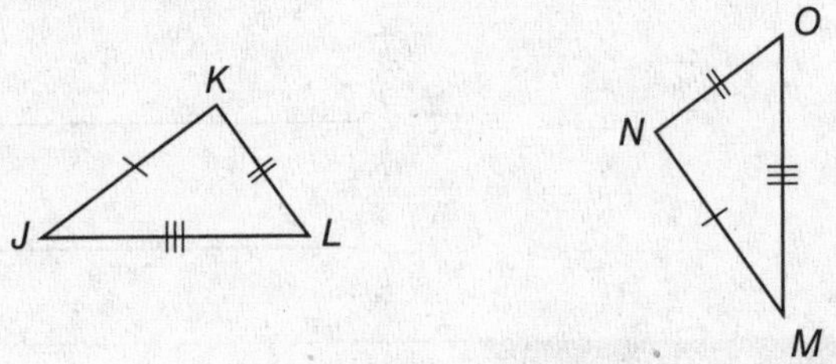

1. $\triangle JKL \cong \triangle$ ☐
2. $\triangle$ ☐ $\cong \triangle NOM$
3. $\triangle JLK \cong \triangle$ ☐
4. $\triangle KJL \cong \triangle$ ☐
5. $\triangle$ ☐ $\cong \triangle OMN$
6. $\triangle LKJ \cong \triangle$ ☐

Identify Congruent Triangles
p. 600

Fill in the diagram to complete the steps to determine congruent triangles. Use the terms *statement*, *order*, *angles*, *vertices*, and *sides*.

Step 1		Step 2		Step 3
Name the corresponding ________.	→	Name the corresponding ________.	→	Write a congruency ________ so the corresponding ________ are in the same ________.

Helping You Remember

Corresponding is a word used in everyday English as well as in mathematics. Write the definition of *corresponding*. Explain how the English definition can help you remember how *corresponding* is used in mathematics.

____

____

NAME ______________________ DATE ____________ PERIOD ____________

11-3 Rotations

What You'll Learn

Skim the Examples for Lesson 11-3. Predict two things that you will learn about rotations.

1. ____

____

2. ____

____

Active Vocabulary

New Vocabulary Match each definition with the correct term by drawing a line to connect the two.

rotation	the fixed point about which a figure is rotated
center of rotation	when a figure can be rotated less than 360° about its center so that its image matches the original figure
rotational symmetry	a transformation where a figure is turned about a fixed point

Vocabulary Link *Rotational* and *symmetry* are two words used in everyday English. Find the definitions of *rotational* and *symmetry* using a dictionary. List three examples of something that has rotational symmetry.

____

____

____

Lesson 11-3

NAME ______________________ DATE __________ PERIOD __________

Main Idea	Details
Rotations pp. 605–607	**Draw the letter after a 90° counterclockwise rotation around the point.**

Main Idea	Details
Rotational Symmetry p. 607	**Fill in the organizer about rotational symmetry.**

How do you decide if a figure has rotational symmetry?	**How do you find the angle of rotation?**
Draw a figure with a 90° angle of rotation.	**Draw a figure with no rotational symmetry.**

Rotational Symmetry

Helping You Remember

A classmate was absent the day that rotation and rotational symmetry were taught. Provide an explanation of the two concepts.

____

____

____

11-4 Quadrilaterals

What You'll Learn

Skim the lesson. Write two things you already know about quadrilaterals.

1. ______________________________________

____

2. ______________________________________

____

Active Vocabulary

Review Vocabulary Write the definition next to each term. *(Lesson 10-3)*

vertex ▶ ______________________________________

____

segment ▶ ______________________________________

____

New Vocabulary Fill in each blank with the correct term or phrase.

quadrilateral ▶ A quadrilateral is a ______________ figure with ______________ sides and ______________ angles. The segments that form a quadrilateral ______________ only at their ______________.

Vocabulary Link *Quad-* is a prefix used in everyday English as well as in mathematics. Write the meaning of the prefix *quad-*. Write two examples of words used in everyday life that have *quad-* as a prefix and their meanings.

____

____

____

Lesson 11-4 *(continued)*

Main Idea	Details
Find Angle Measures pp. 612–613	**Explain how the model proves that a quadrilateral has angles whose measures have a sum of 360°.**

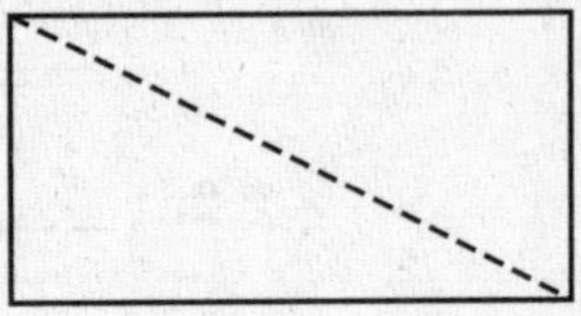

____

____

____

Classify Quadrilaterals p. 613	**Fill in the organizer to classify and describe each figure. Then draw lines to connect the figures and show their relationships.**

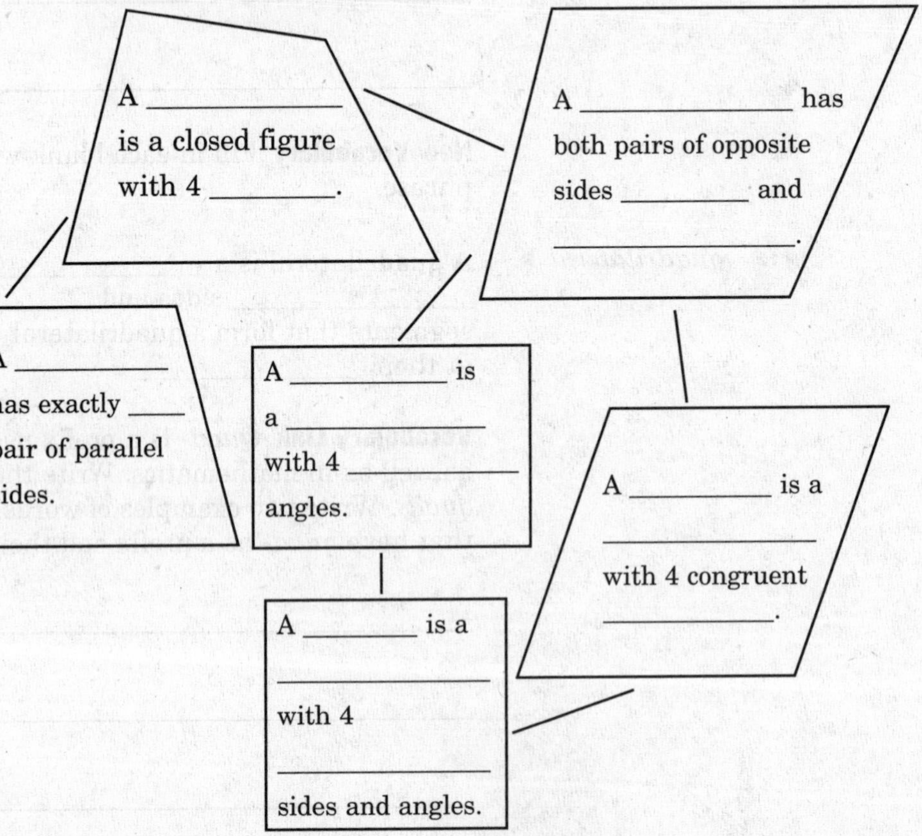

A ________ is a closed figure with 4 ________.
A ________ has both pairs of opposite sides ________ and ________.
A ________ has exactly ____ pair of parallel sides.
A ________ is a ________ with 4 ________ angles.
A ________ is a ________ with 4 congruent ________.
A ________ is a ________ with 4 ________ sides and angles.

NAME ______________________ DATE __________ PERIOD __________

11-5 Polygons

What You'll Learn

Scan the text in Lesson 11-5. Write two facts you learned about polygons.

1. ______________________________________

____

2. ______________________________________

____

Active Vocabulary

New Vocabulary Label the diagram with the correct terms.

diagonal ▶

interior angles ▶

regular polygon ▶

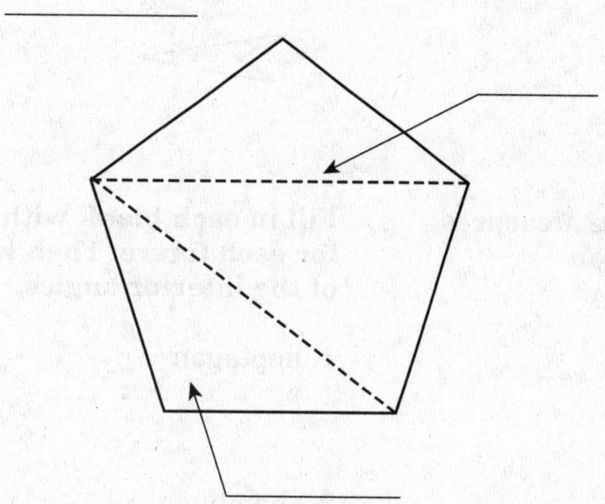

Vocabulary Link *Tessellation* can be illustrated by real-world examples. Look around the room. Give two examples of real-world tessellations.

____

____

____

Lesson 11-5

Main Idea	Details

Classify Polygons
p. 617

Circle the figures that are *not* polygons. If a figure is not a polygon, write the reason inside or beside the figure.

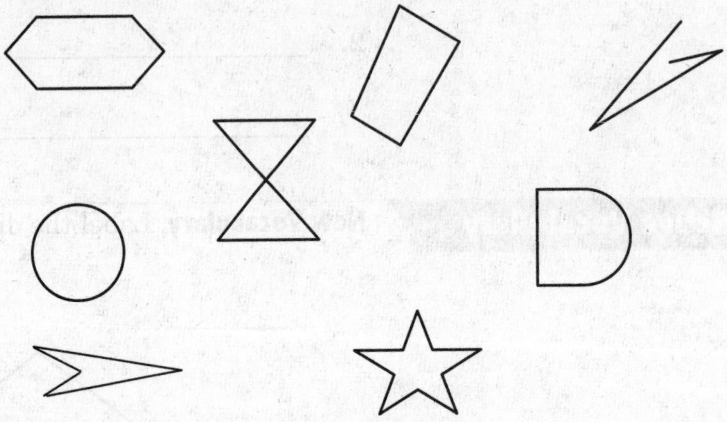

Find Angle Measures of a Polygon
pp. 618–619

Fill in each blank with the number of interior angles for each figure. Then write the sum of the measures of the interior angles.

1. heptagon

   A heptagon is a __-sided figure.
   So, $(n - 2)180 = (\_\_ - 2)180$
   $n =$ _____. The sum is _____.

2. rhombus

   A rhombus is a 4-sided figure.
   So, $(n - 2)180 = (\_\_ - 2)180$
   $n =$ _____. The sum is _____.

Tessellations
p. 619

Cross out the figure that can not be used to make a tessellation. Explain.

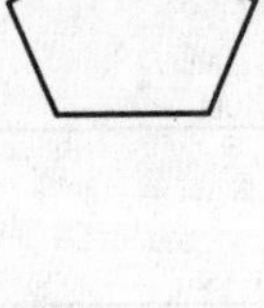

____
____
____
____
____

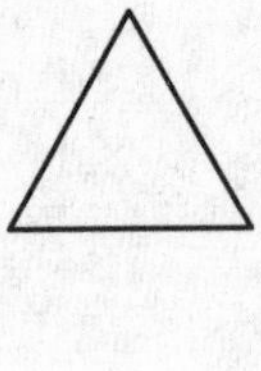

____
____
____
____

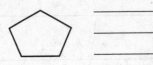

NAME ______________________ DATE __________ PERIOD __________

11-6 Area: Parallelograms, Triangles, and Trapezoids

What You'll Learn

Scan the text under the *Now* heading. List two things you will learn about in the lesson.

1. ____

____

2. ____

____

Active Vocabulary

New Vocabulary Label the diagram with the correct terms.

base ▶

altitude ▶

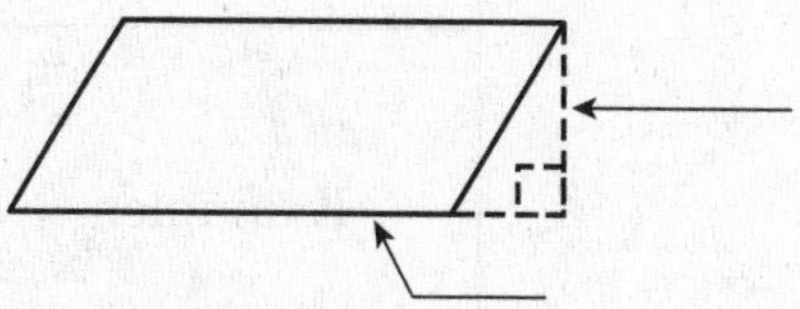

Vocabulary Link *Altitude* is a word that is used in everyday English. Find the definition of *altitude* using a dictionary. Explain how the English definition can help you remember how *altitude* is used in mathematics.

____

____

____

____

____

____

Lesson 11-6

Main Idea	**Details**

Area of Parallelograms
p. 624

Compare the area of a rectangle and the area of a parallelogram.

	Area	
	Rectangle	**Parallelogram**
Formula	$A = lw$	$A =$ ________
Words	Area is length times width.	Area is ________ times ________.
Model	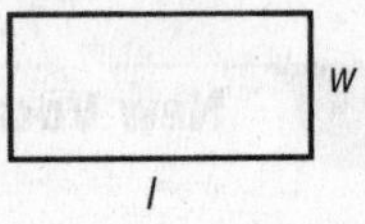	
Examples	A rectangle with length 6 cm and width 5 cm has an area of ________.	A parallelogram with base 10 mm and height 9 mm has an area of ________.

Area of Triangles and Trapezoids
pp. 625–626

Complete to summarize the area of a triangle and a trapezoid.

A parallelogram divided in half by a diagonal results in two congruent triangles. The area of a parallelogram is the sum of the area of the two ________. Because the area of a parallelogram is ________ times ________, the area of a triangle is half the ________ times ________ or ________.

A trapezoid with base a and base b can be divided in half by a ________ resulting in two noncongruent triangles. The sum of the area of those two triangles is $\frac{1}{2}ah + \frac{1}{2}bh$ which is equal to ___(___ + ___).

Helping You Remember

Match the formula with the correct figure by drawing a line to connect them. Then find its area.

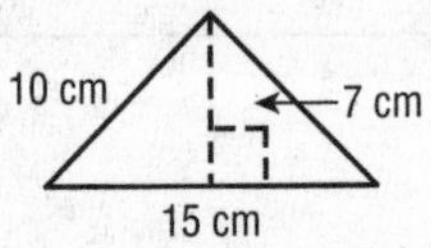

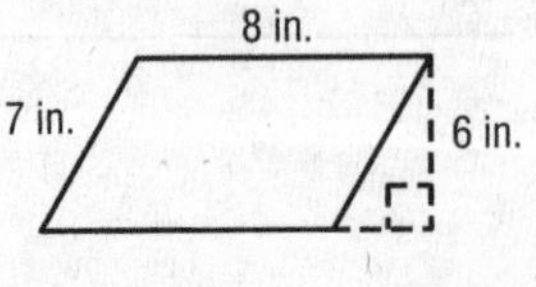

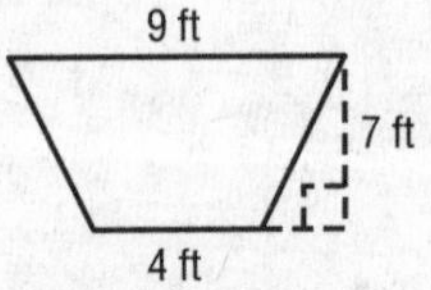

$A = \frac{1}{2}h(a + b)$ $A = \frac{1}{2}(bh)$ $A = bh$

NAME ______________________ DATE __________ PERIOD __________

11-7 Circles and Circumference

What You'll Learn

Skim Lesson 11-7. Predict two things that you expect to learn based on the headings and the Key Concept box.

1. ____

____

2. ____

____

Active Vocabulary

New Vocabulary Fill in each blank with the correct term or phrase.

circle ▶ the set of all ______________ in a plane that is the same ______________ from a given ______________ in the plane

center ▶ the given ______________ in the middle of a ______________

radius ▶ the ______________ from the ______________ to any point on the circle

diameter ▶ the ______________ across the circle through the ______________

chord ▶ the ______________ between any two ______________ on the circle

circumference ▶ the distance ______________ the circle

π (pi) ▶ the ______________ of the ______________ to the ______________ of a circle

Lesson 11-7

Main Idea	Details

Circumference of Circles
pp. 631–632

Label each part of the circle. Then find its circumference with the given diameter or radius. Round to the nearest tenth.

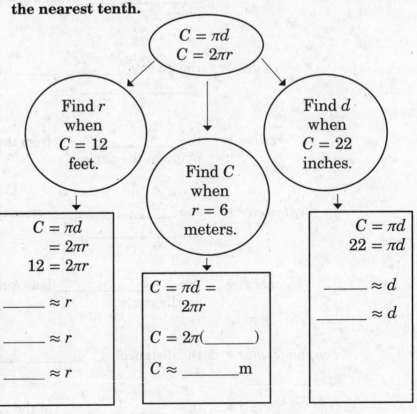

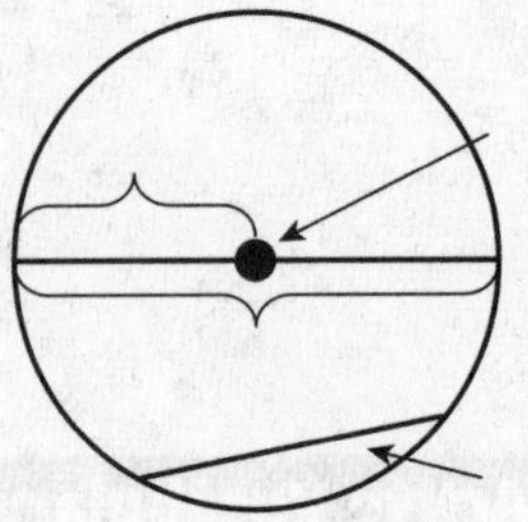

1. $r = 15$ mm Use $C = 2\pi r$.
$C = 2\pi($______$)$
$C =$ ______ π
$C \approx$ ______ mm

2. $d = 8$ yd Use $C = \pi d$.
$C = \pi($______$)$
$C \approx$ ______ yd

Use Circumference to Solve Problems
p. 632

Fill in the blanks to complete the organizer. Round to the nearest tenth.

$C = \pi d$
$C = 2\pi r$

Find r when $C = 12$ feet.
$C = \pi d$
$= 2\pi r$
$12 = 2\pi r$
______ $\approx r$
______ $\approx r$
______ $\approx r$

Find C when $r = 6$ meters.
$C = \pi d = 2\pi r$
$C = 2\pi($______$)$
$C \approx$ ______ m

Find d when $C = 22$ inches.
$C = \pi d$
$22 = \pi d$
______ $\approx d$
______ $\approx d$

NAME ________________ DATE ________ PERIOD ________

11-8 Area of Circles

What You'll Learn

Scan Lesson 11-8. List two headings you would use to make an outline of this lesson.

1. ______________________________

____

2. ______________________________

____

Active Vocabulary

New Vocabulary Label the diagram with the correct term or phrase.

sector ▶

central angle ▶

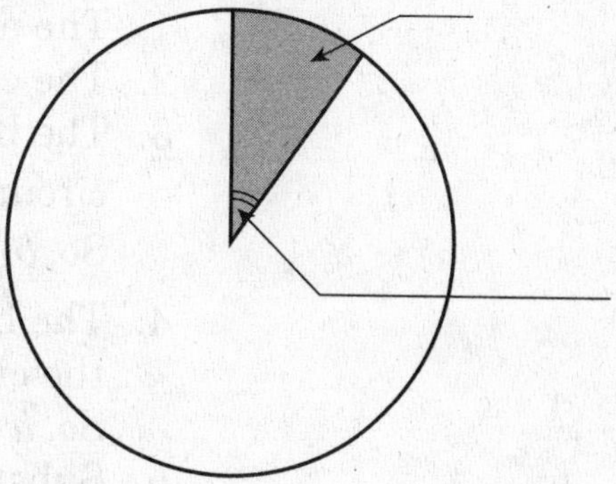

Vocabulary Link *Sector* is a word that is used in everyday English. Find the definition of *sector* using a dictionary. Explain how the English definition can help you remember how *sector* is used in mathematics.

____

____

____

____

____

____

Main Idea	Details

Area of Circles
pp. 636–637

Fill in each blank to summarize the formula for the area of a circle.

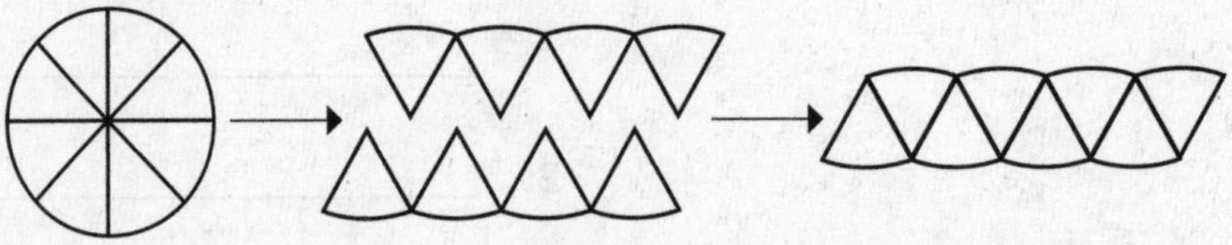

Divide a circle into equal ________.	→	Fit the pieces together to make a ________.

1. The area of a parallelogram is ________.
2. The circumference of a circle is ________.
3. The base of the parallelogram is ________ the circumference of the circle.
   So, $b = \frac{1}{2}C = \frac{1}{2}(\_\_\_\_\_\_) = \_\_\_\_\_\_$.
4. The height of the parallelogram is the ________ of the circle.
   So, $h = \_\_\_\_\_\_$.
5. Substitute the values for b and h.
   $A = bh = \_\_\_\_\_\_ = \_\_\_\_\_\_$

Area of Sectors
p. 638

Fill in each blank to find the area of a sector.

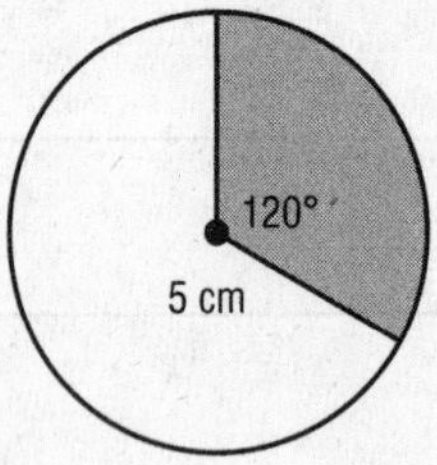

$A = \frac{N}{360}(\pi r^2)$ Use the formula.

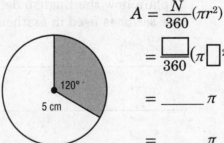

$= \frac{\square}{360}(\pi \square^2)$ N is the number of degrees of the central angle.

$= \_\_\_\_\_\_ \pi$

$= \_\_\_\_\_\_ \pi$ Substitute for N and r.

$\approx \_\_\_\_\_\_$ Simplify.

NAME ______________________ DATE __________ PERIOD __________

11-9 Area of Composite Figures

What You'll Learn

Skim the Examples for Lesson 11-9. Predict two things that you will learn about the area of composite figures.

1. ______________________

____

2. ______________________

____

Active Vocabulary

Review Vocabulary Write the definition next to each term. *(Lesson 11-7)*

circle ▶ ______________________

____

radius ▶ ______________________

diameter ▶ ______________________

chord ▶ ______________________

circumference ▶ ______________________

π(pi) ▶ ______________________

____

New Vocabulary Fill in the blank with the correct term or phrase.

composite figure ▶ A composite figure is made up of ______________________.

Lesson 11-9

Main Idea | **Details**

Area of Composite Figures

pp. 642–643

Fill in each blank to find the area of the composite figure.

Step 1:	Step 2:	Step 3:
Divide the figure into ________ that help you find the area. →	Calculate the ______ of each ________. →	The _____ of the ______ of the ________ is the area of the ___________ figure.

Name the shapes that make up each composite figure. Then draw lines that show the shapes.

1.

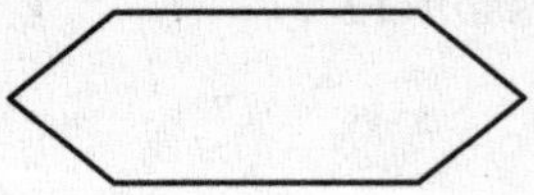

2.

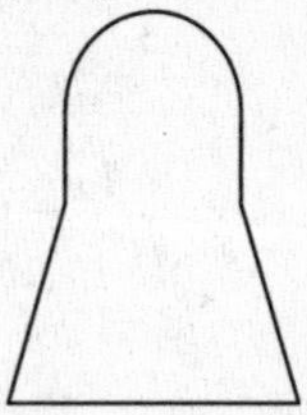

______________________________ ____________________________

____

Helping You Remember

Find the area of each polygon and then the area of the composite figure.

Polygon 1: __________

Polygon 2: __________

Polygon 3: __________

Composite Figure: __________

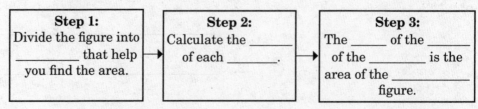

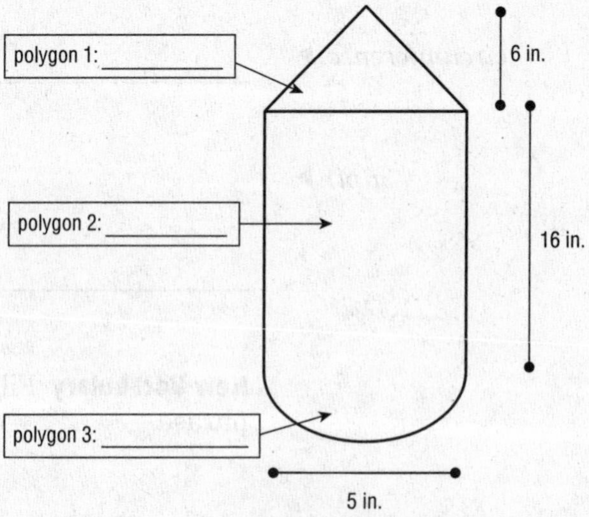

NAME ______________________ DATE ____________ PERIOD ____________

Chapter 11 Distance and Angle

Tie It Together

Complete each graphic organizer with a term or formula from the chapter.

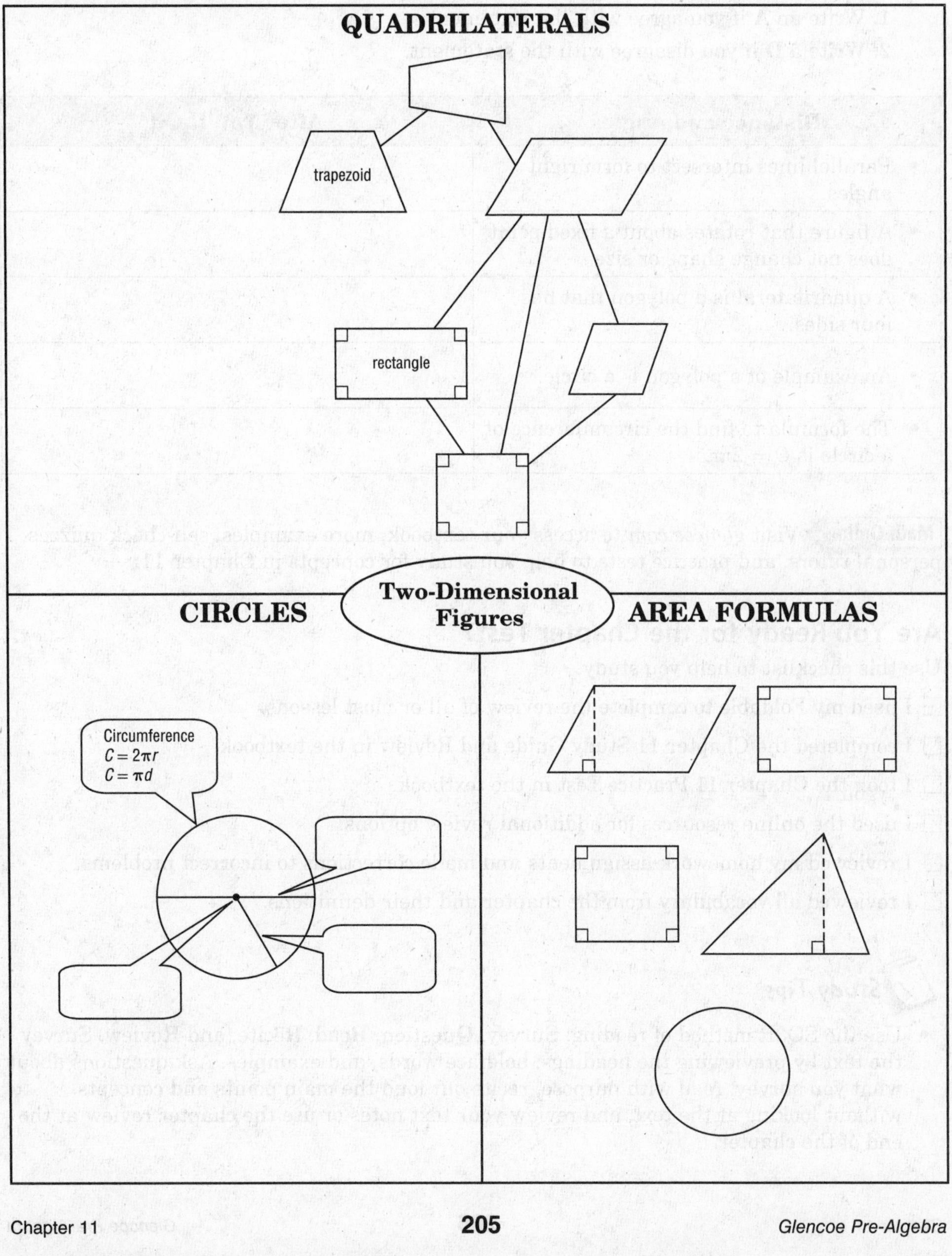

NAME ______________________ DATE __________ PERIOD __________

CHAPTER 11 Distance and Angle

Before the Test

Now that you have read and worked through the chapter, think about what you have learned and complete the table below. Compare your previous answers with these.

1. Write an **A** if you agree with the statement.
2. Write a **D** if you disagree with the statement.

Distance and Angles	After You Read
• Parallel lines intersect to form right angles.	
• A figure that rotates about a fixed point does not change shape or size.	
• A quadrilateral is a polygon that has four sides.	
• An example of a polygon is a circle.	
• The formula to find the circumference of a circle is $C = 2\pi r$.	

Math Online Visit *glencoe.com* to access your textbook, more examples, self-check quizzes, personal tutors, and practice tests to help you study for concepts in Chapter 11.

Are You Ready for the Chapter Test?

Use this checklist to help you study.

☐ I used my Foldable to complete the review of all or most lessons.

☐ I completed the Chapter 11 Study Guide and Review in the textbook.

☐ I took the Chapter 11 Practice Test in the textbook.

☐ I used the online resources for additional review options.

☐ I reviewed my homework assignments and made corrections to incorrect problems.

☐ I reviewed all vocabulary from the chapter and their definitions.

Study Tips

- Use the SQ3R method of reading: **S**urvey, **Q**uestion, **R**ead, **R**ecite, and **R**eview. Survey the text by previewing the headings, boldface words, and examples. Ask questions about what you survey, read with purpose, recite out loud the main points and concepts without looking at the text, and review your text notes or use the chapter review at the end of the chapter.

NAME ______________________ DATE ____________ PERIOD ____________

CHAPTER 12 Surface Area and Volume

Before You Read

Before you read the chapter, think about what you know about surface area and volume. List three things you already know about them in the first column. Then list three things you would like to learn about them in the second column.

K What I know...	W What I want to find out...

FOLDABLES® Study Organizer Construct the Foldable as directed at the beginning of this chapter.

Note Taking Tips

- **Include pictures with your notes.**
  Having diagrams that are labeled with specific parts of each figure will help you understand the formulas.
- **Remember to study your notes daily.**
  Reviewing small amounts at a time will help you retain the information.

NAME ______________________ DATE ____________ PERIOD ____________

CHAPTER 12

Surface Area and Volume

Key Points

Scan the pages in the chapter and write at least one specific fact concerning each lesson. For example, in the lesson on three-dimensional figures, one fact might be that a face is a flat surface. After completing the chapter, you can use this table to review for your chapter test.

Lesson	Fact
12-1 Three-Dimensional Figures	
12-2 Volume of Prisms	
12-3 Volume of Cylinders	
12-4 Volume of Pyramids, Cones, and Spheres	
12-5 Surface Area of Prisms	
12-6 Surface Area of Cylinders	
12-7 Surface Area of Pyramids and Cones	
12-8 Similar Solids	

NAME ______________________ DATE ____________ PERIOD ____________

12-1 Three-Dimensional Figures

What You'll Learn

Skim Lesson 12-1. Predict two things that you expect to learn based on the headings and the Key Concept box.

1. ______________________

2. ______________________

Active Vocabulary

New Vocabulary Write the definition next to each term.

plane ▶ ______________________

____

solid ▶ ______________________

polyhedron ▶ ______________________

edge ▶ ______________________

vertex ▶ ______________________

face ▶ ______________________

prism ▶ ______________________

base ▶ ______________________

pyramid ▶ ______________________

cylinder ▶ ______________________

cone ▶ ______________________

____

cross section ▶ ______________________

Main Idea | **Details**

Identify Three-Dimensional Figures
pp. 664–665

Complete the organizer about *three-dimensional* figures.

What are they?	**Draw a picture of a three-dimensional figure.**
Name some examples.	**Identify these figures.** 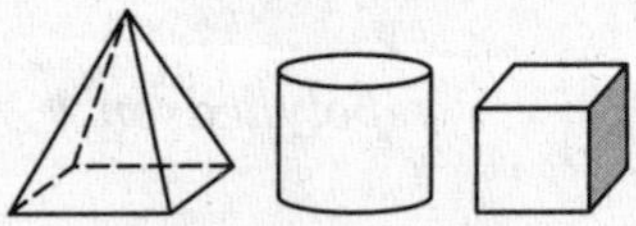

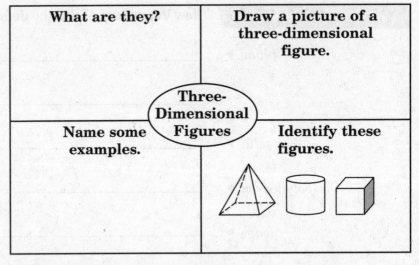

Three-Dimensional Figures

Cross Sections
p. 666

Fill in each blank to summarize cross sections.

1. If a cylinder is sliced vertically, the cross section that would result is a(n) __________.

2. When a triangular pyramid is sliced horizontally, the cross section that results is a(n) __________.

3. When a cone is sliced at an angle, the cross section that results is a(n) __________.

4. If a __________ is sliced vertically, the cross section that would result is a square.

Helping You Remember

The word *polyhedron* is composed of the prefix poly- and the root word-*hedron*. Find the definitions of *poly*- and *hedron*- in a dictionary. Write their definitions.

____

____

12-2 Volume of Prisms

What You'll Learn

Scan Lesson 12-2. List two headings you would use to make an outline of this lesson.

1. ______________________________

____

2. ______________________________

____

Active Vocabulary

Review Vocabulary Match the term with its definition by drawing a line to connect the two. *(Lessons 4-1, 4-2, and 5-1)*

simplest form	equation that shows a relationship between quantities
like terms	terms that contain the same variable
simplifying the expressions	an algebraic expression that has no like terms and no parentheses
formula	combine like terms

New Vocabulary Fill in each blank with the correct word or phrase.

volume ▶ The measure of the ______ occupied by a three-dimensional ______.

Vocabulary Link *Volume* is a word that is used in everyday English. Find the definition of *volume* using a dictionary. Write two sentences of how the word *volume* is used in everyday life.

____

____

____

Lesson 12-2

Main Idea	**Details**

Volume of Prism
pp. 671–672

Compare finding the volume of a rectangular prism with a triangular prism.

	Rectangular Prism	**Triangular Prism**
Formula	$V =$	$V =$
Area of base	$B =$	$B =$
Sketch the prism and find its volume.		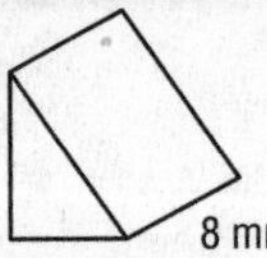

Volume of Composite Figures
p. 673

Find the volume of the figure. Show your work.

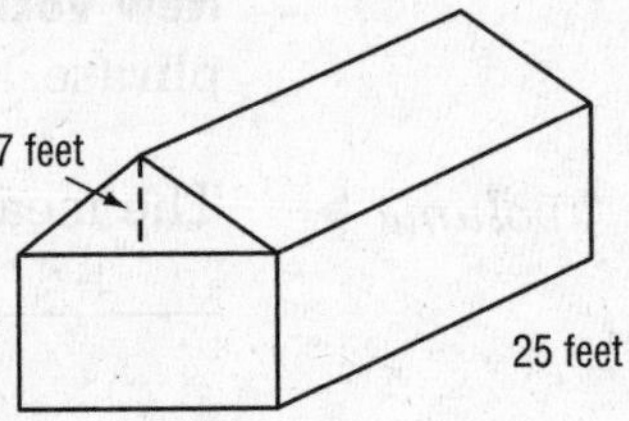

Find the volume of the top. ______________________

Find the volume of the bottom. ______________________

Add the volumes. ______________________

NAME ______________________ DATE __________ PERIOD __________

12-3 Volume of Cylinders

What You'll Learn

Scan the text under the *Now* heading. List two things you will learn about in the lesson.

1. ______________________________

____

2. ______________________________

____

Active Vocabulary

Review Vocabulary Write the term next to each definition. *(Lessons 5-1 and 11-7)*

______________ ▶ the distance around a figure

______________ ▶ an equation that shows a relationship between quantities

______________ ▶ the surface enclosed by a figure

______________ ▶ a set of all points in a plane that is the same distance from a given point called the center

______________ ▶ the given point from which all points on the circle are the same distance

______________ ▶ the distance from the center to any point on the circle

______________ ▶ the distance across the circle through its center

______________ ▶ the distance around a circle

______________ ▶ the ratio of the circumference to the diameter of the circle

Lesson 12-3

NAME ______________________________ DATE ____________ PERIOD ____________

Main Idea | **Details**

Volume of Cylinders
pp. 677–678

Compare how to find the volume of the two figures by completing the chart.

Volume		
	Rectangular Prism	**Cylinder**
Formula	$V = Bh = lwh$	$V = Bh =$
Words	Volume is the area of the base times the height.	Volume is the ____________ times the ________.
Model	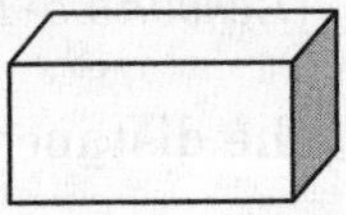	sample model: 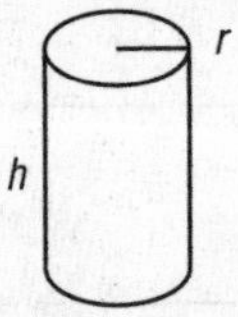
Examples	A rectangular prism with length 5 in., a width 9 in., and a height of 10 in. has a volume of ________.	A cylinder with radius 7 mm and height 15 mm has a volume of ____________.

Volumes of Composite Figures
p. 678

Fill in each blank to complete the steps to find the volume of a composite figure.

Step 1	Step 2	Step 3
________ the figure into ________________.	Find the ________ of each ______ using the correct ________.	Find the ______ of the ________ for the volume of the ________________.

Helping You Remember

Describe how to find the height of a cylinder that has a volume of 2,211 mm^3 and a radius of 8 mm.

____

____

____

NAME ______________________ DATE ____________ PERIOD ____________

12-4 Volume of Pyramids, Cones, and Spheres

What You'll Learn

Skim Lesson 12-4. Predict two things you expect to learn based on the headings and the Key Concept boxes.

1. ______________________

____

2. ______________________

____

Active Vocabulary

Review Vocabulary Match each term with its definition by drawing a line to connect the two. *(Lesson 12-1)*

solid	a three-dimensional figure with one circular base and a vertex connected by a curved side
polyhedron	a three-dimensional figure
prism	a solid with flat surfaces that are polygons
cone	a polyhedron with one base that is a polygon
pyramid	a polyhedron with two parallel congruent bases

New Vocabulary Fill in each blank with the correct term or phrase.

sphere ▶ A set of __________ in space that are a given __________ *r* from the __________.

Lesson 12-4

Main Idea | **Details**

Volume of a Cone
p. 684

Compare the volume of a cylinder and a cone.

Step 1: Find the volume of the two figures.

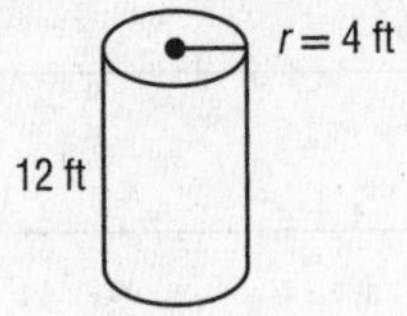

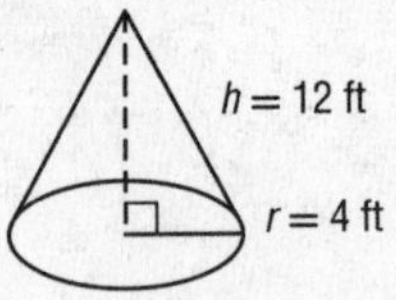

$V =$ $\qquad$ $V =$

$V \approx$ $\qquad$ $V \approx$

Step 2: Make a conjecture about the relationship between the volume of a cylinder and the volume of a cone with the same height and radius.

____

____

Volume of a Sphere
pp. 684–685

Write out each step to find the volume of a sphere with r = 3 cm.

[ ] → formula for the volume of a sphere

[ ] → Substitue 3 for r.

[ ] → Solve for v.

Helping You Remember

Describe each step in finding the volume of the pyramid at the right.

$V = \frac{1}{3}Bh$ ______________________________

$V = \frac{1}{3}(\frac{1}{2} \cdot 5 \cdot 8)27$ ______________________________

$V \approx 180$ ______________________________

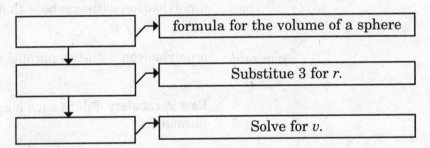

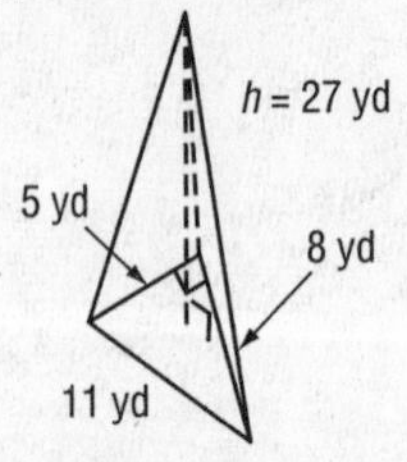

NAME ______________________________ DATE ____________ PERIOD ____________

12-5 Surface Area of Prisms

What You'll Learn

Scan the text in Lesson 12-5. Write two facts you learned about the surface area of prisms as you scanned the text.

1. ____

____

2. ____

____

Active Vocabulary

Review Vocabulary Write the term next to each definition. *(Lessons 5-1 and 12-1)*

____________ ▶ the distance around a figure

____________ ▶ a flat surface of a solid

____________ ▶ one of two congruent parallel faces of a prism

New Vocabulary Write the definition next to each term.

lateral face ▶ ____

lateral area ▶ ____

surface area ▶ ____

Vocabulary Link *Lateral* is a word that is used in everyday English. Find the definition of *lateral* using a dictionary. Explain how the English definition can help you remember how *lateral* is used in mathematics.

____

____

____

Lesson 12-5

Main Idea	Details

Prisms
pp. 691–692

Summarize information about *lateral* and *surface area* in the graphic organizer below.

lateral area → **What is lateral area?** → **How do you find the lateral area?** → **Formulas:**

lateral area: L = ________

surface area: S = ________

surface area → **What is surface area?** → **How do you find the surface area?** → **Formulas:**

Describe how to find the surface area of the figure below.

____

____

____

____

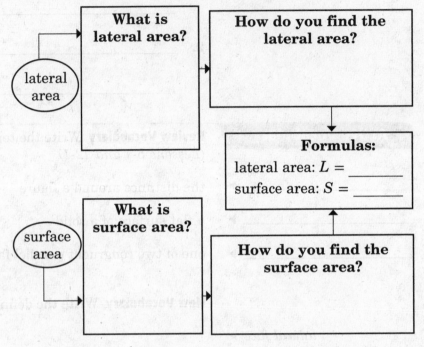

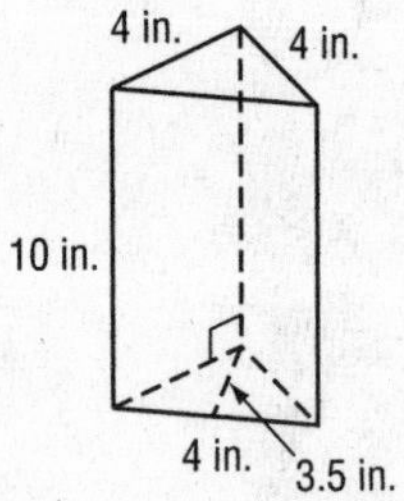

Helping You Remember

How does drawing a net help you find the surface area of a prism? Draw a prism and its net to justify your answer.

____

____

NAME ______________________ DATE __________ PERIOD __________

12-6 Surface Area of Cylinders

What You'll Learn

Skim the lesson. Write two things you already know about the surface area of cylinders.

1. ______________________________

____

2. ______________________________

____

Active Vocabulary

Review Vocabulary Fill in each blank with the correct term or phrase. *(Lesson 12-1)*

cylinder ▶ a three-dimensional figure with congruent, __________ bases that are circles connected by a __________ side

cone ▶ a three-dimensional figure with one __________ base and a __________ connected by a __________ side

vertex ▶ where __________ or more planes intersect at a __________

face ▶ a flat __________ of a __________

Vocabulary Link Cylinders are used in everyday life. List four examples of when the lateral area or surface area of a cylinder may be needed.

____

____

____

____

Main Idea	Details

Surface Area of Cylinders

pp. 697–698

Draw the net of the cylinder. Label the radius (*r*), height (*h*), and circumference (*C*) on the net.

Figure

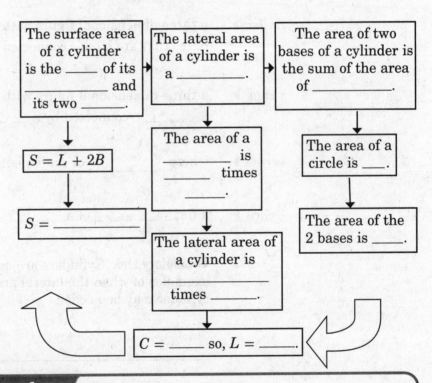

Fill in the blanks to complete the organizer about surface area of a cylinder.

The surface area of a cylinder is the ______ of its ____________ and its two ______.

→ The lateral area of a cylinder is a ____________.

→ The area of two bases of a cylinder is the sum of the area of ____________.

$S = L + 2B$

$S =$ ____________

The area of a ____________ is ______ times ______.

The lateral area of a cylinder is ____________ times ______.

The area of a circle is ____.

The area of the 2 bases is ____.

$C =$ ______ so, $L =$ ______.

Helping You Remember

You want to decorate the side and bottom of a cylindrical flower pot with material. Do you need to calculate the surface area of the pot or the lateral area?

____

____

____

NAME ______________________ DATE __________ PERIOD __________

12-7 Surface Area of Pyramids and Cones

What You'll Learn

Skim the Examples for Lesson 12-7. Predict two things you think you will learn about the surface area of pyramids and cones.

1. ______________________________

____

2. ______________________________

____

Active Vocabulary

Review Vocabulary Write the definition next to each term. *(Lesson 12-5)*

lateral face ▶ ______________________________

____

lateral area ▶ ______________________________

____

surface area ▶ ______________________________

____

New Vocabulary Draw an arrow to the diagram that points the slant height of the pyramid. Then label it with the term *slant height.*

slant height ▶

regular pyramid ▶

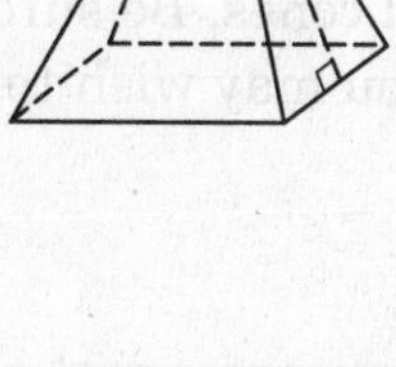

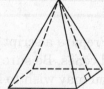

This figure is a __________ __________ because it has a base that is a regular polygon.

Lesson 12-7

Main Idea	Details

Surface Area of Pyramids
pp. 702–703

Draw the net of the pyramid. Label slant height (ℓ), base (B), and side length (s) of the base on the net.

Figure

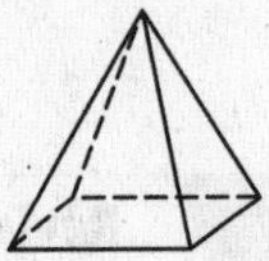

Surface Area of Cones
p. 704

Compare the volume of a cone and the surface area of a cone by filling out the chart. Sample answers are given.

Cone	Volume	Surface Area
Formula	$V = \frac{1}{3}\pi r^2 h$	$S = L + \pi r^2$
Words		
Example	Find the volume of a cone with a radius of 5 cm and a height of 7 cm. $V \approx$ ______	Find the surface area of a cone with a radius of 5 cm and a slant height of 7 cm. $S \approx$ ______

Helping You Remember

Prepare a script for a short presentation on how to find the surface areas of pyramids and cones. Be sure to include any necessary vocabulary terms in your explanation. You may wish to include diagrams with your presentation.

NAME ______________________ DATE __________ PERIOD __________

12-8 Similar Solids

What You'll Learn

Scan Lesson 12-8. List two headings you would use to make an outline of this lesson.

1. ______________________________

____

2. ______________________________

____

Active Vocabulary

Review Vocabulary Match the term with the definition by drawing a line to connect the two. *(Lessons 6-5 and 6-7)*

cross products	an equation that states that two ratios or rates are equal
similar figures	If $\frac{a}{b} = \frac{c}{d}$, then $ad = cb$.
proportion	figures with congruent corresponding angles and proportional corresponding side lengths

New Vocabulary Fill in each blank with the correct term or phrase.

similar solid ▶ Two figures that have the same __________ and their __________ measures are __________.

Vocabulary Link Similar solids are seen in everyday life. Give an example of items that have the same shape but not necessarily the same size in real life.

____

____

____

____

Lesson 12-8

Main Idea	Details

Identify Similar Solids
pp. 709–710

Complete the organizer by filling in each blank to identify similar solids. Then complete the example.

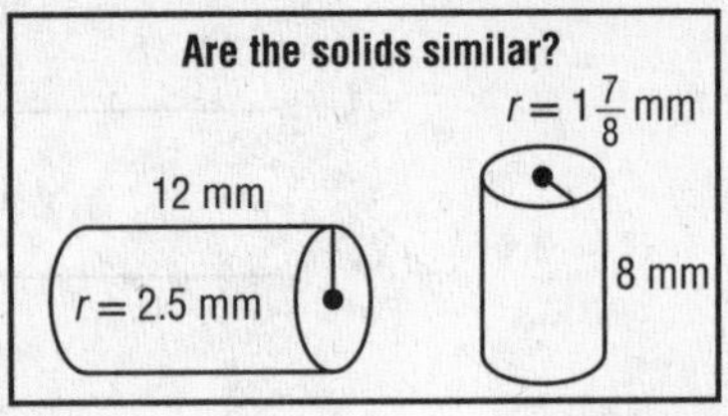

Step 1:
Set up a ____________.

⇩

Step 2:
Find the ______________.

⇩

Step 3:
________.
If the ____________ are ______, the solids are ________.

Step 1:
$\frac{\_\_\_}{12} = \frac{1.875}{\_\_\_}$

⇩

Step 2:
______ = 12(1.875)

⇩

Step3:
______ ≠ 22.5
____, the solids are ______ ______.

Properties of Similar Solids
pp. 710–711

For each pair of solids listed in the table below, describe measurements you would need to determine if the pair is similar.

Pair of Solids	Measurements Needed
Rectangular Prisms	
Cylinders	
Square Pyramids	
Triangular Prisms	
Cones	

Helping You Remember

Describe the relationship between similar figures for surface area and volume.

____

____

NAME ______________________ DATE ____________ PERIOD ________

CHAPTER 12 Surface Area and Volume

Tie It Together

Fill in the formulas for each solid. Label the appropriate variables on each figure.

	Prism	Cylinder	Pyramid
Volume			
Lateral Area			
Surface Area			
Diagram/Variables	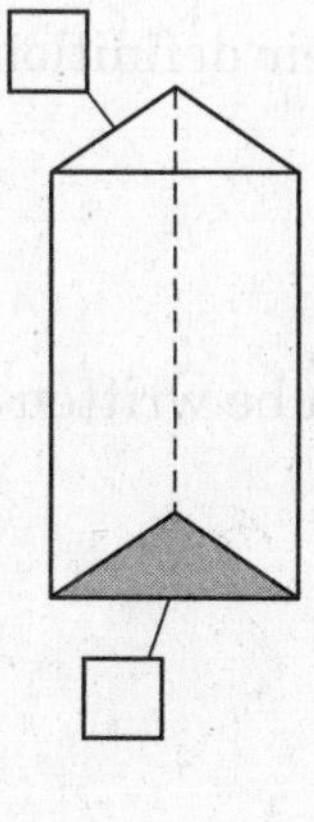	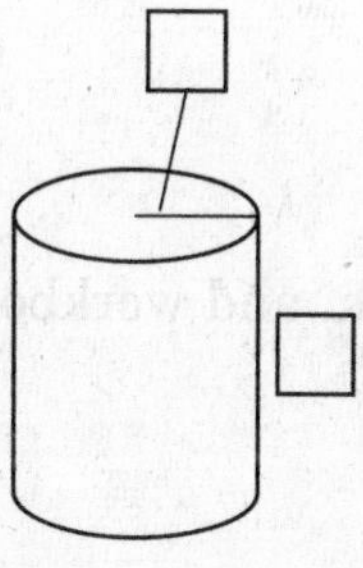	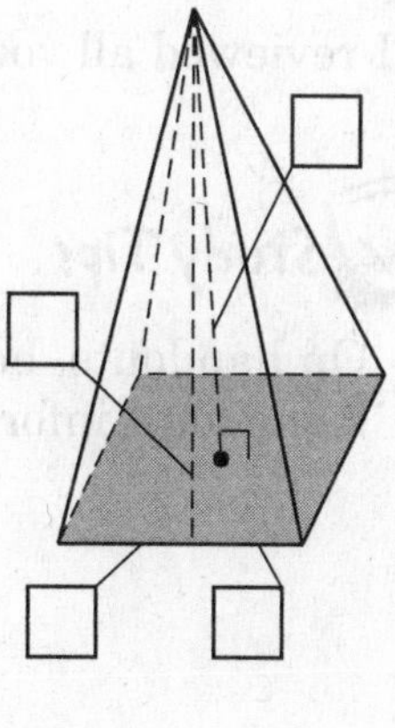

NAME ______________________________ DATE ____________ PERIOD ____________

CHAPTER 12 Surface Area and Volume

Before the Test

Review the ideas you listed in the table at the beginning of the chapter. Cross out any incorrect information in the first column. Then complete the table by filling in the third column.

K What I know...	W What I want to find out...	L What I learned...

Math Online Visit *glencoe.com* to access your textbook, more examples, self-check quizzes, personal tutors, and practice tests to help you study for concepts in Chapter 12.

Are You Ready for the Chapter Test?

Use this checklist to help you study.

☐ I used my Foldable to complete the review of all or most lessons.

☐ I completed the Chapter 12 Study Guide and Review in the textbook.

☐ I took the Chapter 12 Practice Test in the textbook.

☐ I used the online resources for additional review options.

☐ I reviewed my homework assignments and made corrections to incorrect problems.

☐ I reviewed all vocabulary from the chapter and their definitions.

Study Tips

- On handouts, homework, and workbooks that can be written in, underline and highlight significant information.

NAME ______________________ DATE ____________ PERIOD ____________

CHAPTER 13 Statistics and Probability

Before You Read

Before you read the chapter, respond to these statements.

1. Write an **A** if you agree with the statement.
2. Write a **D** if you disagree with the statement.

Before You Read	Statistics and Probability
	• The median of a set of data is the same thing as the average.
	• The range is the difference between the least and greatest numbers.
	• A histogram is a type of graph that uses bars.
	• Probability is expressed as a number between 1 and 100.
	• When something is likely to happen, it is certain.

FOLDABLES® Study Organizer Construct the Foldable as directed at the beginning of this chapter.

Note Taking Tips

- **When you take notes, it may be helpful to sit as close as possible to the front of the class.**
  There are fewer distractions and it is easier to hear.
- **When taking notes on statistics, include your own statistical examples as you write down concepts and definitions.**
  This will help you to better understand statistics.

NAME ______________________________ DATE ____________ PERIOD ________

CHAPTER 13 Statistics and Probability

Key Points

Scan the pages in the chapter and write at least one specific fact concerning each lesson. For example, in the lesson on measures of variation, one fact might be that the median of a set of data separates the set in half. After completing the chapter, you can use this table to review for your chapter test.

Lesson	Fact
13-1 Measures of Central Tendency	
13-2 Stem-and-Leaf Plots	
13-3 Measures of Variation	
13-4 Box-and-Whisker Plots	
13-5 Histograms	
13-6 Theoretical and Experimental Probability	
13-7 Using Sampling to Predict	
13-8 Counting Outcomes	
13-9 Permutations and Combinations	
13-10 Probability of Compound Events	

NAME ______________________________ DATE ____________ PERIOD ____________

13-1 Measures of Central Tendency

What You'll Learn

Skim Lesson 13-1. Predict two things that you expect to learn based on the headings and the Key Concept box.

1. ______________________________

____

2. ______________________________

____

Active Vocabulary

New Vocabulary Fill in each blank with the missing term or phrase.

mode ▶ the ________ or ________ that occur ________

median ▶ the ____________ when data is ordered from ________ to ________ or the ________ of the ________ two numbers

measures of central tendency ▶ describes the ____________ of the data

mean ▶ the ____________ of the data ____________ by the ____________ of items in the ____________ set

Vocabulary Link *Median* is a word that is used in everyday English. Find the definition of *median* using a dictionary. Give two examples of how *median* might be used in everyday life.

____

____

____

Lesson 13-1

Main Idea	Details
Measures of Central Tendency pp. 730–732	**Complete the organizer. Write the three kinds of measures of central tendency with its definition. Then write a problem with the solution to show an example for each.**

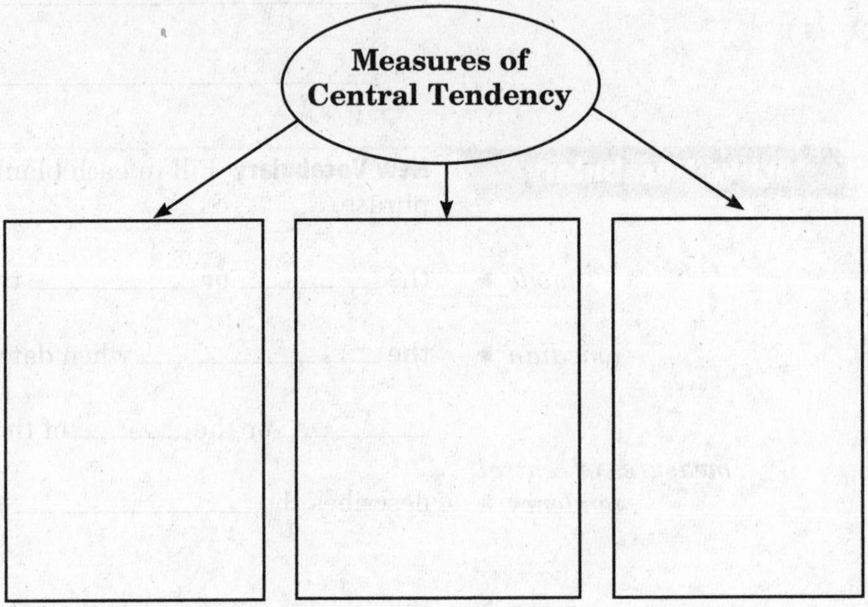

The heights of a group of friends are 54, 62, 48, 62, 58, and 58 inches. Fill in each blank to find the measures of central tendency.

1. Write the numbers in order from ________ to __________.

2. Find the _________ of the numbers and ________ by ______.

3. The mean is _________, the median is _________, and the mode is __________.

NAME ______________________ DATE ____________ PERIOD ____________

13-2 Stem-and-Leaf Plots

What You'll Learn

Scan Lesson 13-2. List two headings you would use to make an outline of this lesson.

1. ______________________________________

____

2. ______________________________________

____

Active Vocabulary

New Vocabulary Label the *stems* and *leaves* in both plots. Then name the two types of plots.

stems ▶

leaves ▶

stem-and-leaf plot ▶

back-to-back stem-and-plot ▶

8 6 3	0	4 7 9
0	1	3 7 7 9
8 5 3 3	2	0 0 5 8
9 9 7 4	3	0 0 1 1 3
2 1	4	3 7
	5	1

3|1 = 31

0	4 7 9
1	3 7 7 9
2	0 0 5 8
3	0 0 1 1 3 4 7
4	3 7
5	1

0|9 = 9

____

Main Idea

Details

Display Data
p. 737

Explain the steps to construct a stem-and-leaf plot.

Step 1:
Find the ________ and ________ numbers. Identify the ________ place value. The ________ are the greatest place value of the data.

→

Step 2:
List the ________ in order in the stem column. Write the rest of the digits in the ________ column. Order the leaves from ________ to ________.

→

Step 3:
Remember to include a ________ and a ________.

Lesson 13-2

Main Idea | **Details**

Interpret Data
p. 739

Complete the stem-and-leaf plot using the data in the table.

Books Checked Out Weekly					
115	113	125	145	119	117
101	156	154	118	154	132
100	122	106	111	126	130

What does the Stem '14' represent? __________

What is the greatest number of books checked out? __________

What is the mode of the data? __________

How many weeks does the data cover? ______

Helping You Remember

Measures of central tendency can be easily found using a stem-and-leaf plot. Explain how you could use the data in the stem-and-leaf plot below to find the mean, median, and mode. Then find the measures.

13-3 Measures of Variation

What You'll Learn

Skim the Examples for Lesson 13-3. Predict two things that you will learn about measures of variation.

1. ____

____

2. ____

____

Active Vocabulary

Review Vocabulary Write the term next to the definition. *(Lesson 13-1)*

________________ ▶ The middle number when data is ordered from least to greatest or the mean of the middle two numbers

New Vocabulary Fill in the diagram with correct terms. Then complete the statements.

range ▶

quartiles ▶

upper quartile ▶

lower quartile ▶

interquartile range ▶

outlier ▶

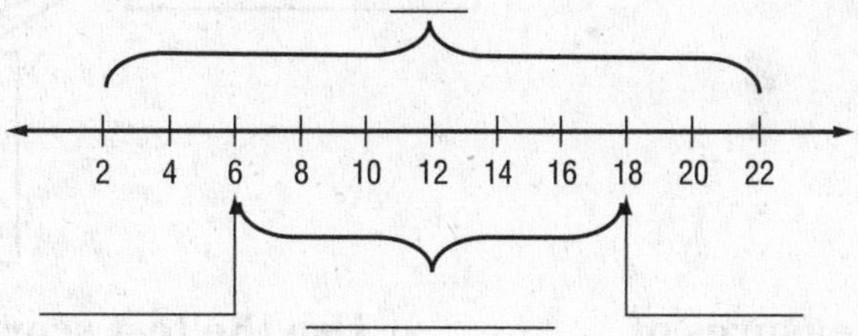

__________ are the values that divide the data into four equal parts.

An __________ is a value that is more than 1.5 times the value of the interquartile range beyond the quartiles.

Lesson 13-3

Main Idea | **Details**

Measures of Variation
pp. 743–745

Complete the diagram and the example.

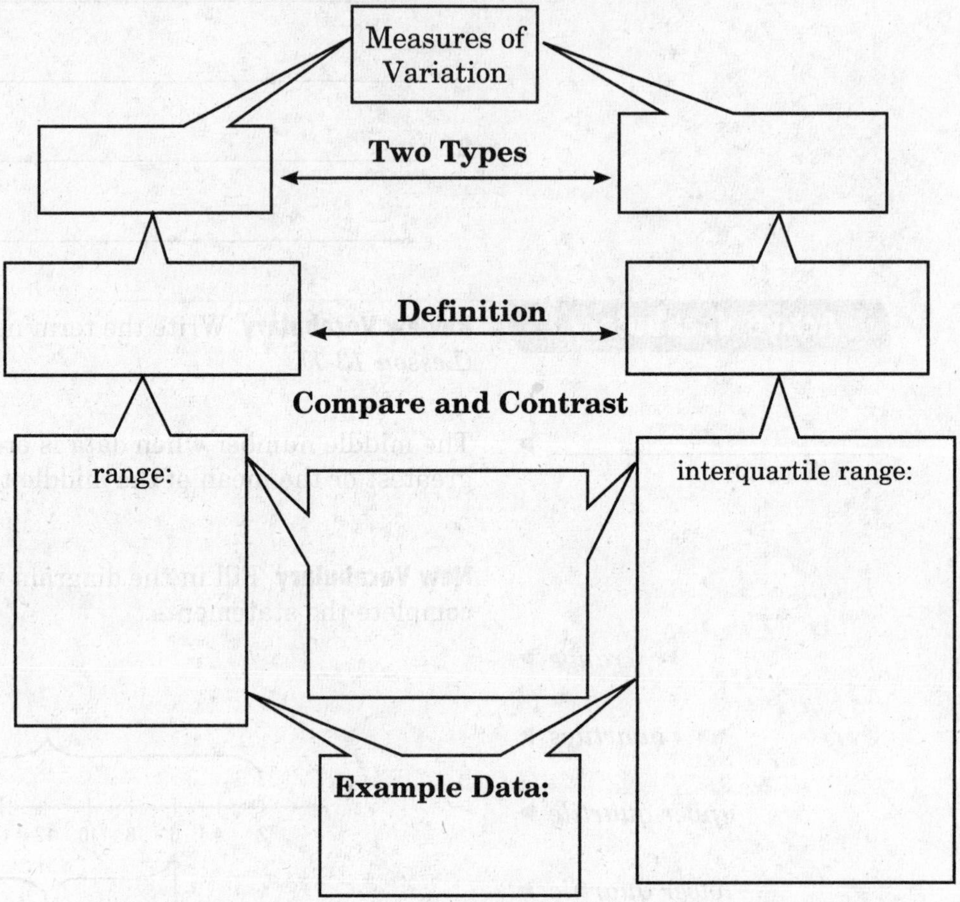

Use Measures of Variation
pp. 745–746

Use the test scores from the table to answer the questions below.

Jackson	67	80	78	75	80	79	77	79	55
Terry	68	77	60	77	71	72	52	63	59

What is the range of Jackson's and Terry's scores? ________

What conclusions can be drawn from the ranges? ________

____

____

What are the interquartile ranges for each student?

____

What conclusions can be drawn from the interquartile ranges? ______________________________

____

NAME ______________________ DATE __________ PERIOD __________

13-4 Box-and-Whisker Plots

What You'll Learn

Skim the lesson. Write two things you already know about box-and-whisker plots.

1. ________________________________

____

2. ________________________________

____

Active Vocabulary

New Vocabulary Fill in each blank with the correct term or phrase.

box-and-whisker plot ▶ uses a __________ to show the __________ of a set of __________; also known as a __________

Main Idea

Details

Display Data
p. 750

Complete the organizer to explain the steps to construct a box-and-whisker plot. Then complete the example.

Step 1: Draw a __________ that includes the __________ and __________ values.	**Data:** 14, 33, 28, 9, 32, 37, 31
↓	↓
Step 2: Mark the __________ and __________ values, the ______, the ______ and ______ __________. Check for ________.	
↓	↓
Step 3: Draw the ______ and ________.	

Lesson 13-4

Main Idea	Details
Interpret Box-and-Whisker Plots pp. 751–752	**Use the information from the box-and-whisker plot to answer each question.**

Ages of Arcade Players

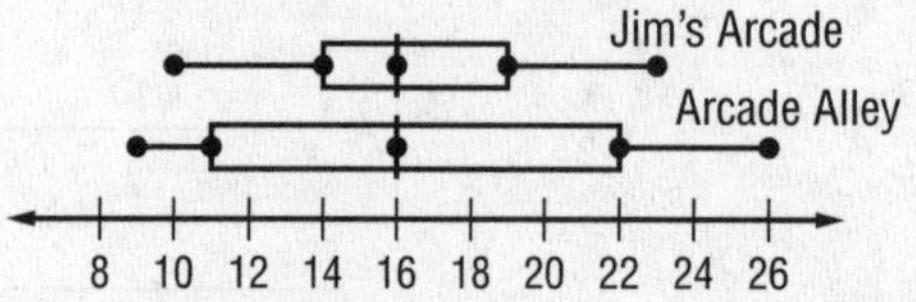

1. Which arcade attracts a wider range of ages? ____________
2. What age is 25% of the age group at Jim's Arcade less than? ____________
3. Compare the median for both arcades. What can you conclude?

____

____

Helping You Remember

Describe in detail how to determine if an outlier exists in a data set.

____

____

____

____

____

____

NAME ______________________ DATE ____________ PERIOD ________

13-5 Histograms

What You'll Learn

Scan the text in Lesson 13-5. Write two facts you learned about histograms.

1. ____

____

2. ____

____

Active Vocabulary

Review Vocabulary Write the term under the correct display. *(Lessons 13-2 and 13-4)*

stem-and-leaf plot ▶

box-and-whisker plot ▶

histogram ▶

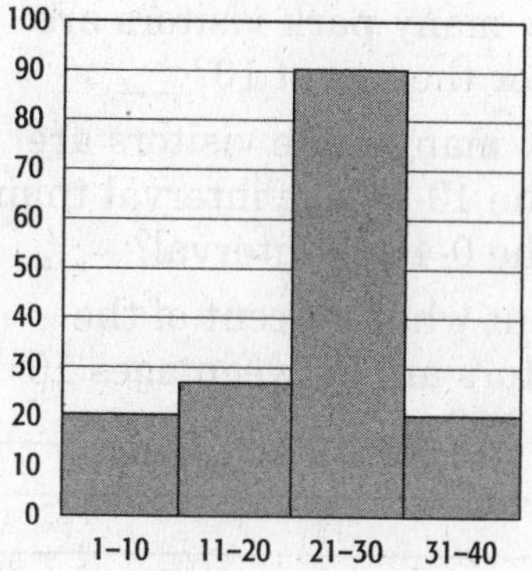

Stem	Leaf
0	4 7 9
1	3 7 7 9
2	0 0 5 8
3	0 3
4	3 7
5	1

3 | 3 = 33

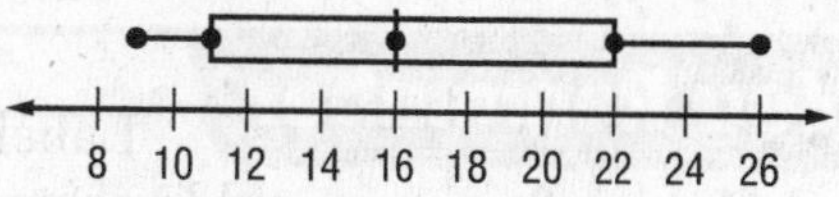

New Vocabulary Write the term *histogram* under the correct display.

Vocabulary Link The data on a histogram is in equal intervals. Name three examples of data that could be displayed in a histogram.

____

Lesson 13-5

Main Idea | **Details**

Displayed Data
p. 757

Cross out the part of the concept circle that does not belong. Explain.

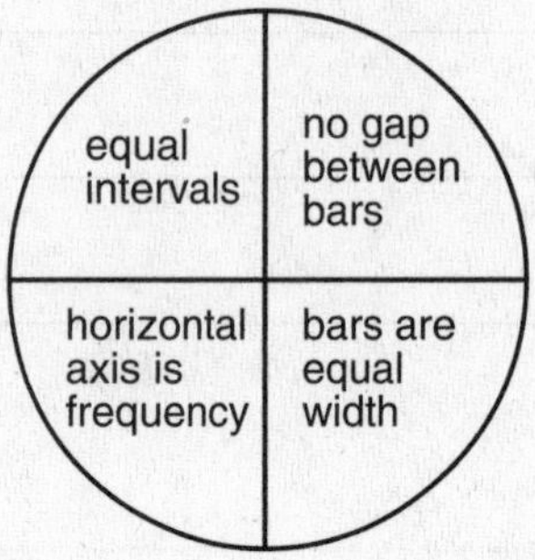

____

____

____

____

Interpret Data
p. 758

Use the information from the histogram to answer each question.

1. How many park visitors are under the age of 10?____
2. How many more visitors are in the 10-14 age interval than in the 0-4 age interval?____
3. About what percent of the visitors are between ages 15 and 19? ____________

____

____

____

____

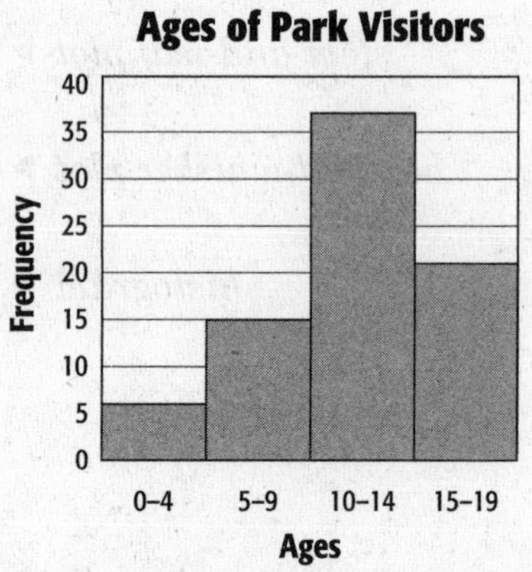

Helping You Remember

Label the histogram: frequency, bar, interval, and histogram. Make a frequency table showing the same information as the histogram.

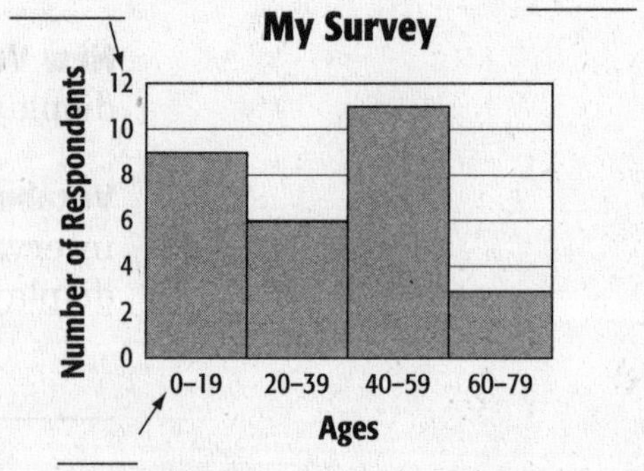

NAME ______________________ DATE __________ PERIOD __________

13-6 Theoretical and Experimental Probability

Lesson 13-6

What You'll Learn

Scan the text under the *Now* heading. List two things you will learn about in the lesson.

1. ____

____

2. ____

____

Active Vocabulary

New Vocabulary Fill in each blank with the correct term or phrase.

simple event ▶ one ________ or a collection of __________

outcomes ▶ the ______ of an event or experiment

sample space ▶ the set of all possible __________

random ▶ when each outcome is equally _____ to occur

probability ▶ a _____ that compares the number of __________ outcomes to the number of ________ outcomes

theoretical probability ▶ what _______ occur in an experiment

experimental probability ▶ what ________ occurs when repeating a probability experiment many times

odds in favor ▶ the _____ that compares the number of ways an event ____ occur to the number of ways that the event _______ occur

odds against ▶ the ____ that compares the number of ways an event _______ occur to the number of ways that the event ____ occur

Main Idea | **Details**

Probability of Simple Events
pp. 765–767

Fill in each blank with the terms, *certain*, *impossible*, or *equally likely*. Then answer the questions below with *true* or *false*.

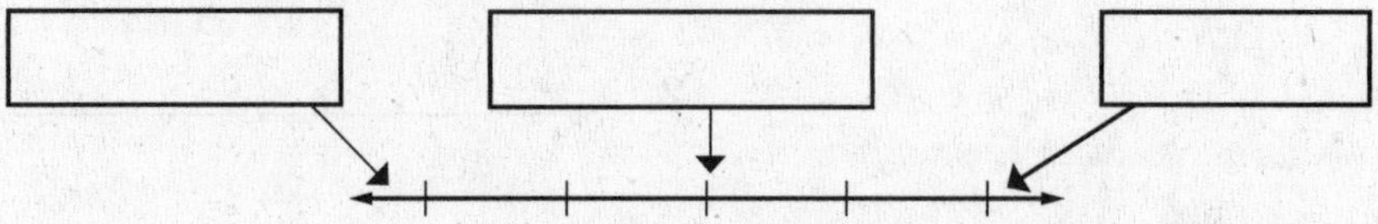

1. Probability is a ratio that compares the number of possible outcomes to the number of favorable incomes. _____
2. The closer a probability is to 1, the less likely it is to occur. _____
3. Experimental probability is what actually happens, while theoretical probability is what should happen. _____

Use a Sample to Make a Prediction
p. 767

The table shows the results from a survey that asked students about their favorite school subject. If 50 more students are picked at random, predict how many will *not* have a favorite subject of math?

Favorite Subject	
subject	**frequency**
science	22
social studies	22
language arts	26
math	30

Helping You Remember

Look up *theoretical* and *experimental* in the dictionary. How can the definitions help you to remember the difference between *theoretical* probability and *experimental* probability?

____

____

____

NAME ______________________ DATE __________ PERIOD __________

13-7 Using Sampling to Predict

What You'll Learn

Skim the lesson. Write two things you already know about using sampling to predict.

1. ______________________

2. ______________________

Active Vocabulary

New Vocabulary Write the definition next to each term.

sample ▶ ______________________

population ▶ ______________________

unbiased sample ▶ ______________________

simple random sample ▶ ______________________

stratified random sample ▶ ______________________

systemic random sample ▶ ______________________

biased sample ▶ ______________________

convenience sample ▶ ______________________

voluntary response sample ▶ ______________________

Lesson 13-7

Main Idea	Details

Identify Sampling Techniques

pp. 771–772

Compare *biased* and *unbiased* sampling techniques by completing the chart below. Sample answers are given.

Technique	Biased Sampling	Unbiased Sampling
What is it?		
How are they the same?		
How are they different?		
What are some examples?		
Is it biased or unbiased?	on-line polls that request visitors to participate: _______ a surveyor who visits every 25th house in neighborhood: _________	

Validating and Predicting Samples

pp. 772–773

A manufacturer makes 1500 phones and tests every 10th phone for defects. Of the phones, 24 were defective.

Is this sampling valid? ____

How many of the 1500 could you expect to be defective? ____

Helping You Remember

How can you remember the difference between *biased* and *unbiased* sampling?

____

____

NAME ______________________________ DATE ____________ PERIOD ____________

13-8 Counting Outcomes

What You'll Learn

Skim Lesson 13-8. Predict two things that you expect to learn based on the headings and the Key Concept box.

1. ______________________________
____

2. ______________________________
____

Active Vocabulary

Review Vocabulary Match each definition with the term by drawing a line to connect the two. (*Lesson 13-6*)

random	the results of an event or experiment
sample space	the set of all possible outcomes
probability	when each outcome is likely to occur
outcomes	a ratio that compares the number of favorable outcomes to the number of possible outcomes

New Vocabulary Fill in each blank with the missing term or phrase.

tree diagram ▶ a ________ that shows different __________ for an ______ or ______________________

Fundamental Counting Principle ▶ ________ the number of __________ to the ________ of ________

Lesson 13-8

Main Idea	Details

Counting Outcomes
pp. 777–778

Write the two methods to find possible outcomes of an event. Then use each method to find the outcomes of the example.

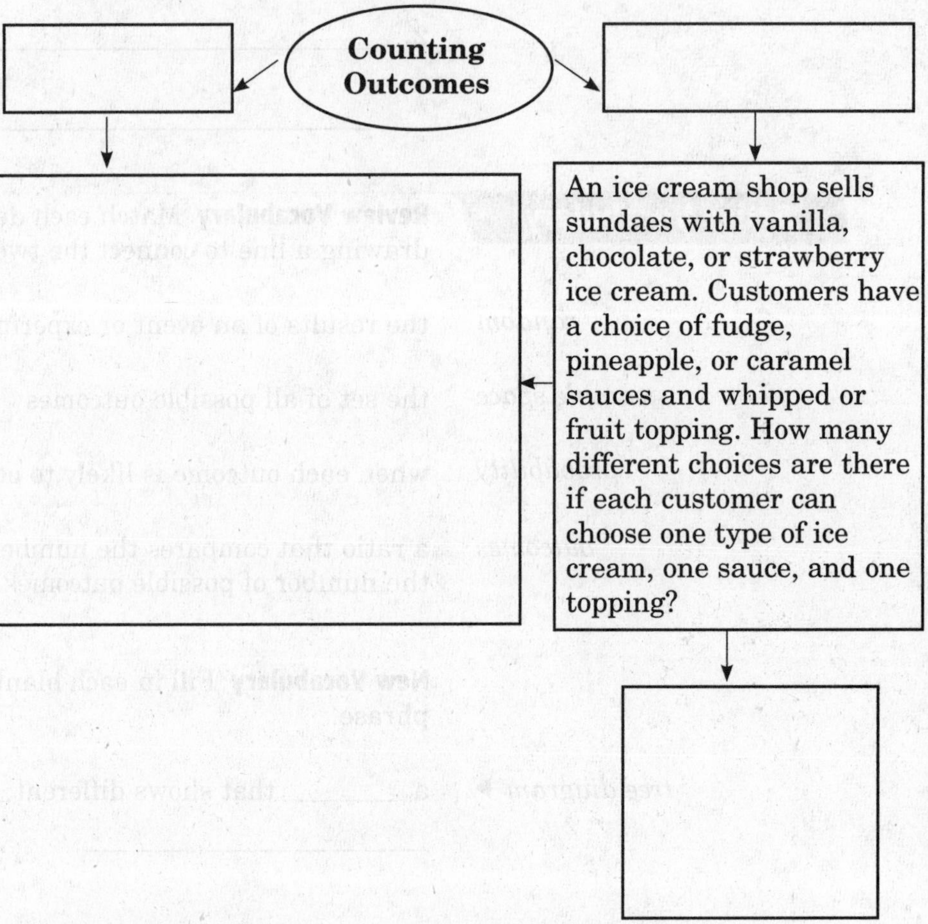

Counting Outcomes
An ice cream shop sells sundaes with vanilla, chocolate, or strawberry ice cream. Customers have a choice of fudge, pineapple, or caramel sauces and whipped or fruit topping. How many different choices are there if each customer can choose one type of ice cream, one sauce, and one topping?

Find the Probability of an Event
pp. 778–779

Find each probability using a number cube labeled 1 through 6.

1. What is the probability of tossing a 1 and then a 2?

2. What is the probability of tossing a number greater than 4 on two consecutive tosses?

NAME ______________________ DATE __________ PERIOD __________

13-9 Permutations and Combinations

What You'll Learn

Scan Lesson 13-9. List two headings you would use to make an outline of this lesson.

1. ______________________________

____

2. ______________________________

____

Active Vocabulary

Review Vocabulary Write the definition next to each term. (*Lesson 13-6*)

random ▶ ______________________________

probability ▶ ______________________________

____

theoretical probability ▶ ______________________________

experimental probability ▶ ______________________________

____

New Vocabulary Fill in each blank with the correct word or phrase.

permutations ▶ an __________ or listing in which order _____ important

combinations ▶ an __________ or listing in which order _____ important

Vocabulary Link The root of permutation is *permute*. Look up *permute* in the dictionary. How can the English definition help you remember the mathematic definition?

____

____

____

Lesson 13-9

Main Idea	Details
Use Permutations pp. 783–784	**Fill in each blank to answer the question.** How many ways can a 4-digit PIN number be made using the numbers 0 through 9 if each number can only be used once? $P(\_\ \_) =$ — Write the notation for a permutation with 10 digits used 4 at a time. $P(\_\ \_) = \_ \cdot \_ \cdot \_ \cdot \_$ $= \_\_\_$ — Use the Fundamental Counting Principle to find the number of possible permutations.
Use Combinations pp. 784–785	**Cross out the part of the concept circle that does not belong. Then state the relationship between the remaining parts.**

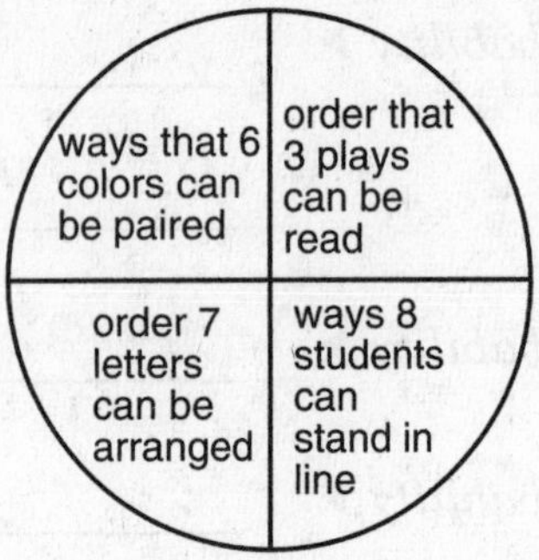

____

____

Helping You Remember

Complete the diagram by writing the words *combinations* and *permutations* in the correct blanks. Then write a sentence based on the diagram stating the difference between permutations and combinations.

____

____

____

____

____

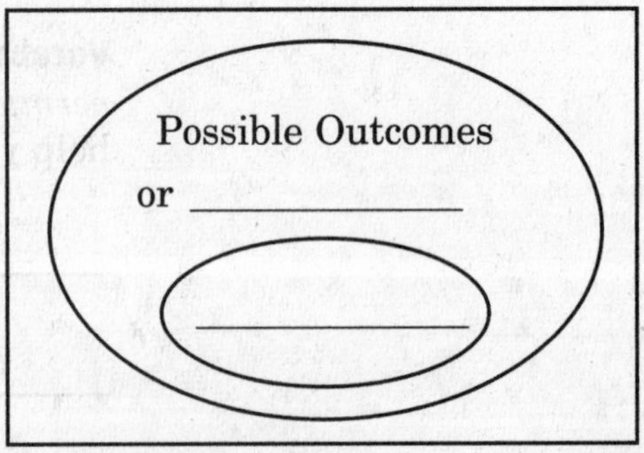

NAME ______________________ DATE ____________ PERIOD ____________

13-10 Probability of Compound Events

What You'll Learn

Skim the Examples for Lesson 13-10. Predict two things that you will learn about the probability of compound events.

1. ______________________________

____

2. ______________________________

____

Active Vocabulary

Review Vocabulary Write the term next to the definition. (*Lesson 13-6*)

______________ ▶ one outcome or a collection of outcomes

New Vocabulary Write the term next to each definition.

______________ ▶ The outcome of one event does not influence the outcomes of a second event.

______________ ▶ consists of two or more simple events

______________ ▶ two events that cannot happen at the same time

______________ ▶ The outcomes of one event affects the outcomes of a second event.

Vocabulary Link *Independent* and *dependent* are words that are used in everyday English. Describe an independent and dependent event that occurs in everyday life.

____

____

____

____

____

Lesson 13-10

NAME ______________________ DATE __________ PERIOD __________

Main Idea	Details

Probabilities of Independent and Dependent Events

pp. 790–791

Fill in each blank with *dependent* or *independent*.

1. A card is turned over and a number cube is tossed.

   ____

2. One marble is randomly picked from a bag. Then a second marble is chosen without replacing the first marble. __________

3. A scarf is randomly chosen from a bag. After putting the first scarf back into the bag, another scarf is chosen.

   ____

4. Two coins are tossed at the same time. __________

5. There are a dozen different flavored bagels in a bag. Jackson reaches in and grabs one. Then Iona grabs one.

Mutually Exclusive Events

p. 792

Compare finding the probability of an independent or dependent event, and two mutually exclusive events.
Sample answers are given.

	Independent Events	Dependent Events	Mutually Exclusive Events
What is it?			
How do you find the probability?			

NAME ______________________ DATE __________ PERIOD __________

CHAPTER 13

Statistics and Probability

Tie It Together

List concepts and vocabulary from the chapter that fit into each square.

Statistics	Data Displays
Sampling	**Probability**

CHAPTER 13

Statistics and Probability

Before the Test

Now that you have read and worked through the chapter, think about what you have learned and complete the table below. Compare your previous answers with these.

1. Write an **A** if you agree with the statement.
2. Write a **D** if you disagree with the statement.

Statistics and Probability	After You Read
• The median of a set of data is the same thing as the average.	
• The range is the difference between the least and greatest numbers.	
• A histogram is a type of graph that uses bars.	
• Probability is expressed as a number between 1 and 100.	
• When something is likely to happen, it is certain.	

Math Online Visit *glencoe.com* to access your textbook, more examples, self-check quizzes, personal tutors, and practice tests to help you study for concepts in Chapter 13.

Are You Ready for the Chapter Test?

Use this checklist to help you study.

☐ I used my Foldable to complete the review of all or most lessons.

☐ I completed the Chapter 13 Study Guide and Review in the textbook.

☐ I took the Chapter 13 Practice Test in the textbook.

☐ I used the online resources for additional review options.

☐ I reviewed my homework assignments and made corrections to incorrect problems.

☐ I reviewed all vocabulary from the chapter and their definitions.

Study Tips

- If possible, rewrite your notes. Not only can you make them clearer and neater, rewriting them will help you remember the information.